BARRON'S

EARLY ACHIEVER

GRADE 1

MATH WORKBOOK
ACTIVITIES & PRACTICE

REVIEW · UNDERSTAND · DISCOVER

Published by Kaplan North America, LLC, d/b/a Barron's Educational Series
1515 W. Cypress Creek Road
Fort Lauderdale, FL 33309
www.barronseduc.com

ISBN 978-1-5062-8135-3

10 9 8 7 6 5 4 3 2 1

Kaplan North America, LLC, d/b/a Barron's Educational Series print books are available at special quantity discounts to use for sales promotions, employee premiums, or educational purposes. For more information or to purchase books, please call the Simon & Schuster special sales department at 866-506-1949.

BARRON'S

Introduction

Barron's Early Achiever workbooks are based on sound educational practices and include both parent-friendly and teacher-friendly explanations of specific learning goals and how students can achieve them through fun and interesting activities and practice. This exciting series mirrors the way mathematics is taught in the classroom. Early Achiever Grade 1 Math presents these skills through different units of related materials that reinforce each learning goal in a meaningful way. The Review, Understand, and Discover sections assist parents, teachers, and tutors in helping students apply skills at a higher level. Additionally, students will become familiar and comfortable with the manner of presentation and learning, as this is what they experience every day in the classroom. These factors will help early achievers master the skills and learning goals in math and will also provide an opportunity for parents to play a larger role in their children's education.

Introduction to Problem-Solving and Mathematical Practices

This book will help to equip both students and parents with strategies to solve math problems successfully. Problem solving in the mathematics classroom involves more than calculations alone. It involves a student's ability to consistently show his or her reasoning and comprehension skills to model and explain what he or she has been taught. These skills will form the basis for future success in meeting life's

goals. Working through these skills each year through the twelfth grade sets the necessary foundation for collegiate and career success. Your student will be better prepared to handle the challenges that await him or her as he or she gradually enters into the global marketplace.

Making Sense of the Problem-Solving Process

For students: It is important that you be able to make sense of word problems, write word problems with numbers and symbols, and be able to prove when you are right as well as to know when a mistake happened. You may solve a problem by drawing a model, using a chart, list, or other tool. When you get your correct answer, you must be able to explain how and why you chose to solve it that way. Every word problem in this workbook allows you to practice these skills, helping to prepare you for the demands of problem solving in your first-grade classroom. The first unit of this book discusses the **Ace It Time!** section of each lesson. **Ace It Time!** will help you master these skills.

While Doing Mathematics You Will...

1. Make sense of problems and become a champion in solving them

- Solve problems and discuss how you solved them
- Look for a starting point and plan to solve the problem
- Make sense (meaning) of a problem and search for solutions

- Use concrete objects or pictures to solve problems
- Check over work by asking, "Does this make sense?"
- Plan out a problem-solving approach

2. Reason on concepts and understand that they are measurable

- Understand numbers represent specific quantities
- Connect quantities to written symbols
- Take a word problem and represent it with numbers and symbols

- Know and use different properties of operations
- Connect addition and subtraction to length

3. Construct productive arguments and compare the reasoning of others

- Construct arguments using concrete objects, pictures, drawings, and actions
- Practice having conversations/discussions about math
- Explain your own thinking to others and respond to the thinking of others

- Ask questions to clarify the thinking of others (How did you get that answer? Why is that true?)
- Justify your answer and determine if the thinking of others is correct

4. Model with mathematics

- Determine ways to represent the problem mathematically
- Represent story problems in different ways; examples may include numbers, words, drawing pictures, using objects, acting out, making a chart or list, writing equations
- Make connections between different representations and explain
- Evaluate your answers and think about whether or not they make sense

5. Use appropriate tools strategically

- Consider available tools when solving math problems
- Choose tools appropriately
- Determine when certain tools might be helpful
- Use technology to help with understanding

6. Attend to detail

- Develop math communication skills by using clear and exact language in your math conversations
- Understand meanings of symbols and label appropriately
- Calculate accurately

7. Look for and make use of structure

- Apply general math rules to specific situations
- Look for patterns or structure to help solve problems
- Adopt mental math strategies based on patterns, such as making ten, fact families, and doubles

8. Look for and express regularity in repeated reasoning

- Notice repeated calculations and look for shortcut methods to solve problems (for example, rounding up and adjusting the answer to compensate for the rounding)
- Evaluate your own work by asking, "Does this make sense?"

Contents

Mathematical Foundations for Grade 1

Problem-Solving Concepts

FOLLOWING THE OBJECTIVE
You will learn how to solve word problems.

LEARN IT: This book will teach you a lot of math. You will learn to add, subtract, measure, graph, and describe shapes. There are steps you can use to solve all kinds of problems. You will use these steps in the *Ace It Time!* part of each lesson.

STEP 1: UNDERSTAND

What's the Question?

The checklist shows steps you can use to solve a math problem. The first step is to read the problem. Ask, "What is the question?" Find the question. Underline it. Then check "Yes" on the checklist.

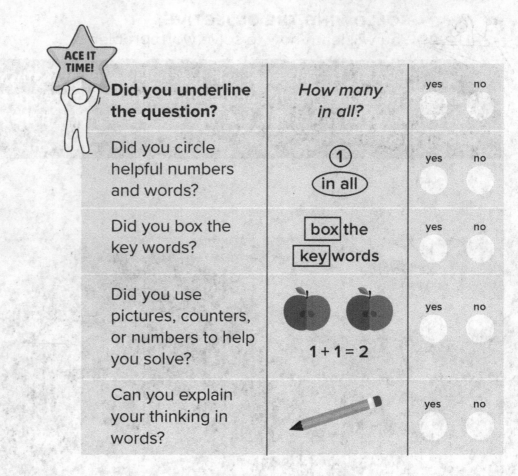

ACE IT TIME!		yes	no
Did you underline the question?	*How many in all?*	○	○
Did you circle helpful numbers and words?	① in all	○	○
Did you box the key words?	box the key words	○	○
Did you use pictures, counters, or numbers to help you solve?	1 + 1 = 2	○	○
Can you explain your thinking in words?		○	○

PRACTICE: Underline the question.

Example: Mark picks 6 apples. Annie picks 3 apples. <u>How many apples did they pick in all?</u>

STEP 2: IDENTIFY THE NUMBERS

What Numbers Are Needed to Solve the Problem?

Find the numbers you need. Circle them. Then check "Yes" on the checklist.

PRACTICE: Circle the numbers you need to solve the problem.

Example: Mark picks ⑥ apples. Annie picks ③ apples. <u>How many apples did they pick in all?</u>
Circle 6 and 3. Those are the numbers of apples.

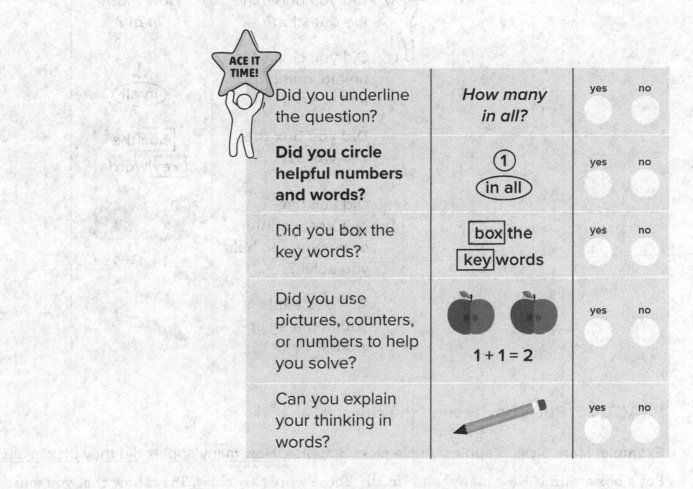

	ACE IT TIME!		yes	no
	Did you underline the question?	*How many in all?*	○	○
	Did you circle helpful numbers and words?	① in all	○	○
	Did you box the key words?	box the key words	○	○
	Did you use pictures, counters, or numbers to help you solve?	1 + 1 = 2	○	○
	Can you explain your thinking in words?		○	○

STEP 3: NAME THE OPERATION

In every problem, there will be clues that help you figure out if you are adding or subtracting. Put a box around the clues. Then check "Yes" on the checklist.

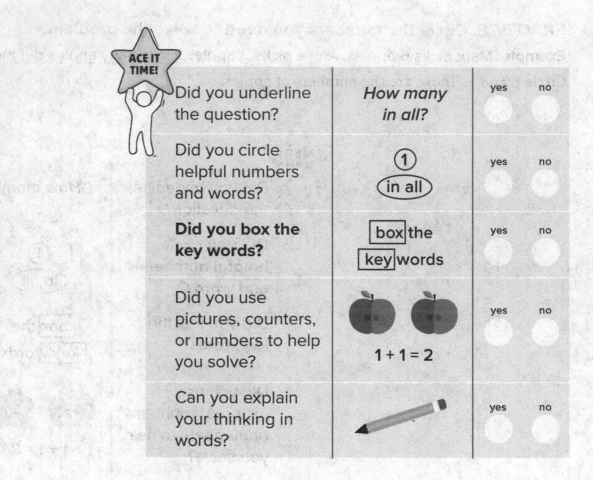

ACE IT TIME!			yes	no
Did you underline the question?	*How many in all?*		○	○
Did you circle helpful numbers and words?	① in all		○	○
Did you box the key words?	box the key words		○	○
Did you use pictures, counters, or numbers to help you solve?	1 + 1 = 2		○	○
Can you explain your thinking in words?			○	○

PRACTICE: Put a box around the clues.

Example: Mark picks ⑥ apples. Annie picks ③ apples. How many apples did they pick in all?

Put a box around "How many" and "in all." Those words are clues. They show that you must find a total. You can add to find a total.

STEP 4: USE A MODEL TO SOLVE

You can draw a picture to show the problem. You can use counters or other math tools to act out the problem. You can write a number sentence. Choose a good way to model the problem. Then check "Yes" on the checklist.

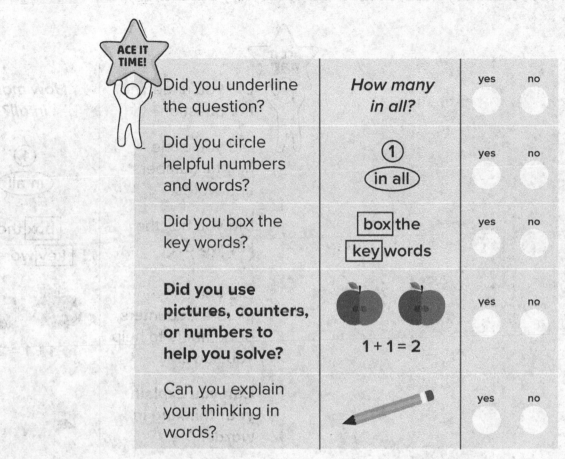

		yes	no
Did you underline the question?	*How many in all?*	○	○
Did you circle helpful numbers and words?	① in all	○	○
Did you box the key words?	box the key words	○	○
Did you use pictures, counters, or numbers to help you solve?	1 + 1 = 2	○	○
Can you explain your thinking in words?		○	○

ACE IT TIME!

PRACTICE: Draw a picture to model the problem.

Label the picture. Write a number sentence.

LET'S PRACTICE:

Example: Mark picks ⑥ apples. Annie picks ③ apples. How many apples did they pick in all?

● ● ● ● ● ●   ● ● ●

Mark   Annie

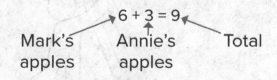

6 + 3 = 9

Mark's apples   Annie's apples   Total

STEP 5: USE MATH VOCABULARY TO EXPLAIN

Write a response. Use math vocabulary.

You are almost done! Explain how you found the answer. Use words, numbers, or pictures. Use the vocabulary words in the Math Vocabulary box to help you!

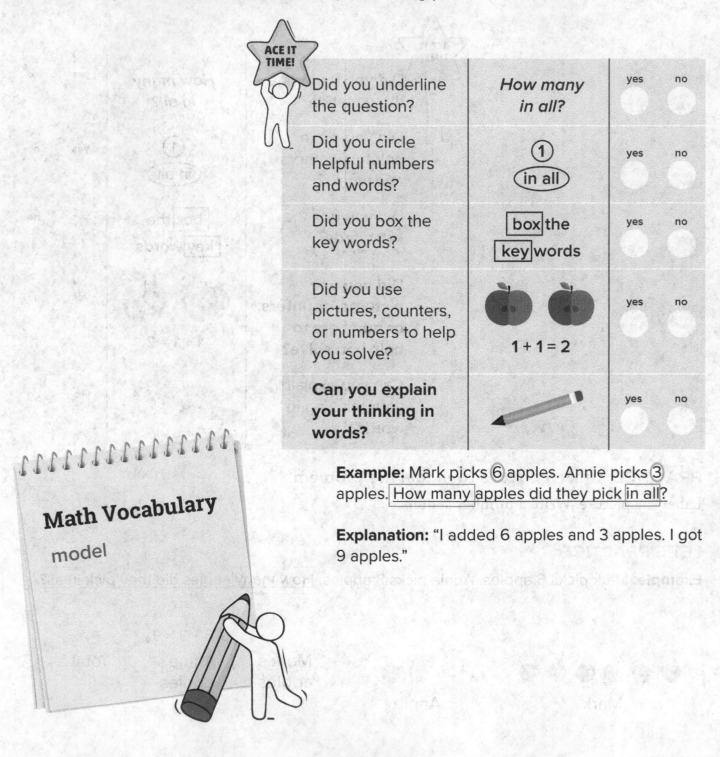

ACE IT TIME!			yes	no
	Did you underline the question?	*How many in all?*	○	○
	Did you circle helpful numbers and words?	① (in all)	○	○
	Did you box the key words?	box the key words	○	○
	Did you use pictures, counters, or numbers to help you solve?	1 + 1 = 2	○	○
	Can you explain your thinking in words?		○	○

Example: Mark picks ⑥ apples. Annie picks ③ apples. How many apples did they pick in all?

Explanation: "I added 6 apples and 3 apples. I got 9 apples."

Math Vocabulary

model

Use Pictures to Add

FOLLOWING THE OBJECTIVE
You will use pictures to show addition.

LEARN IT: You have a group of things. You put more things in the group. *Count* to find how many things in all. That is *addition*. You *add* to find the *sum*.

Example: You have 3 oranges. You get 2 more oranges.

The problem has pictures. Count 3 oranges. Then count 2 more. Write the sum.

3 oranges   plus   2 more oranges =   5 oranges

If there are no pictures, you can draw pictures. Draw circles to show the oranges.

think!
Each circle shows
1 orange.

3 circles   plus   2 more circles

You can also use an *addition sentence* to show the problem and the sum.

think!
You add 3 plus 2. The
sum is 5.

3 + 2 = 5
plus   equals

PRACTICE: Now you try

Use the pictures. Write the sum.

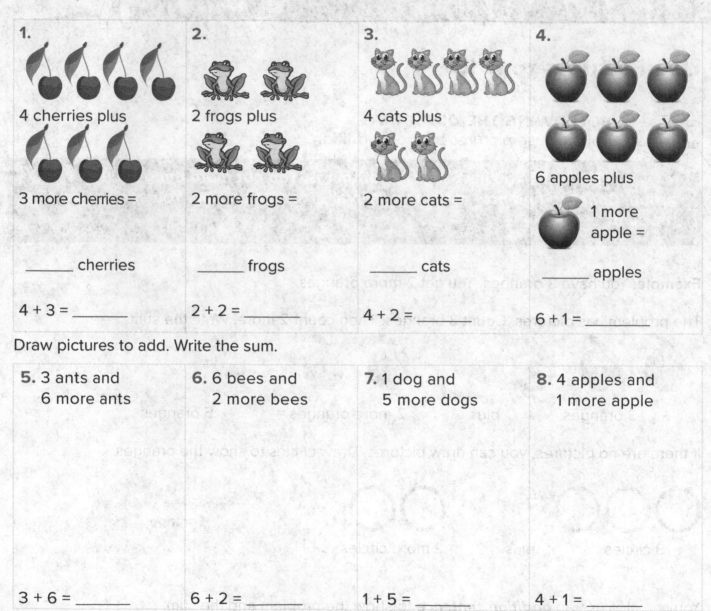

1.

4 cherries plus

3 more cherries =

_____ cherries

4 + 3 = _____

2.

2 frogs plus

2 more frogs =

_____ frogs

2 + 2 = _____

3.

4 cats plus

2 more cats =

_____ cats

4 + 2 = _____

4.

6 apples plus

1 more apple =

_____ apples

6 + 1 = _____

Draw pictures to add. Write the sum.

5. 3 ants and
6 more ants

3 + 6 = _____

6. 6 bees and
2 more bees

6 + 2 = _____

7. 1 dog and
5 more dogs

1 + 5 = _____

8. 4 apples and
1 more apple

4 + 1 = _____

There are 5 birds. 3 more birds come. How many birds are there in all? Show your work and explain your thinking here.

ACE IT TIME!

	How many in all?	yes	no
Did you underline the question?		◯	◯
Did you circle helpful numbers and words?	① in all	◯	◯
Did you box the key words?	box the key words	◯	◯
Did you use pictures, counters, or numbers to help you solve?	$1 + 1 = 2$	◯	◯
Can you explain your thinking in words?		◯	◯

Math Vocabulary

add

addition

addition sentence

equals (=)

plus (+)

sum

Math on the Move

Try adding with objects in your home. Get 10 crayons. Use them to solve this: Sam has 5 crayons. He finds 4 more crayons. How many crayons does he have?

Use Counters to Add

FOLLOWING THE OBJECTIVE
You will use counters to show addition as putting together.

LEARN IT: You can use *counters* to add. Counters are objects you can count. Ask an adult for two types of counters. Try pennies, nickels, or other small objects.

Example: There are 4 red crayons and 2 yellow crayons. How many crayons are there all together?

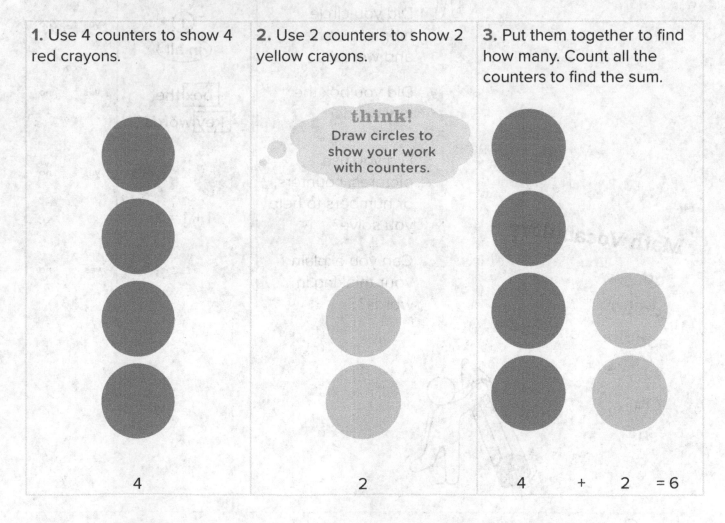

1. Use 4 counters to show 4 red crayons.

2. Use 2 counters to show 2 yellow crayons.

3. Put them together to find how many. Count all the counters to find the sum.

think!
Draw circles to show your work with counters.

4

2

4 + 2 = 6

PRACTICE: Now you try

Use counters to add. On a separate piece of paper, draw circles to show your work. Use 2 colors.

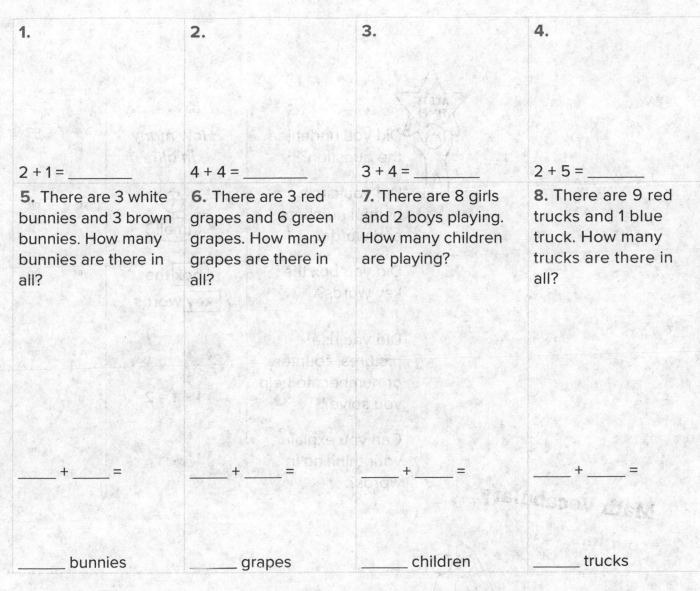

1.	2.	3.	4.
2 + 1 = _____	4 + 4 = _____	3 + 4 = _____	2 + 5 = _____
5. There are 3 white bunnies and 3 brown bunnies. How many bunnies are there in all?	**6.** There are 3 red grapes and 6 green grapes. How many grapes are there in all?	**7.** There are 8 girls and 2 boys playing. How many children are playing?	**8.** There are 9 red trucks and 1 blue truck. How many trucks are there in all?
____ + ____ =	____ + ____ =	____ + ____ =	____ + ____ =
_____ bunnies	_____ grapes	_____ children	_____ trucks

Jan buys 4 red apples and 5 green apples. How many apples does Jan buy? Use counters. Show your work and explain your thinking here.

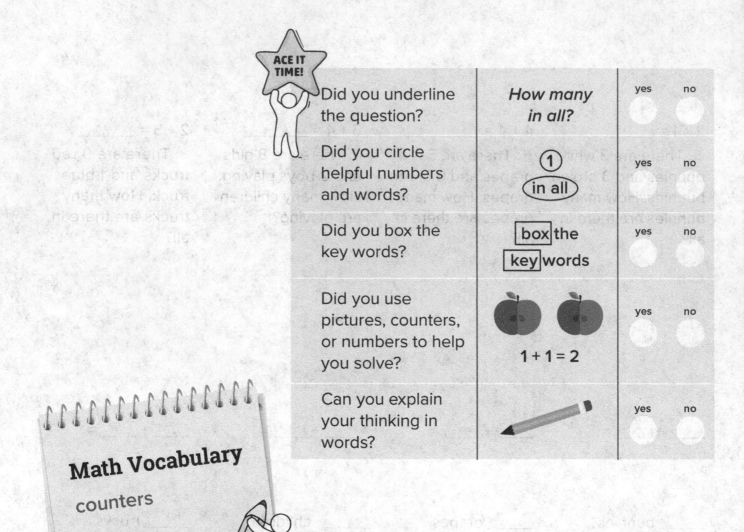

ACE IT TIME!

		yes	no
Did you underline the question?	*How many in all?*	yes	no
Did you circle helpful numbers and words?	① in all	yes	no
Did you box the key words?	box the key words	yes	no
Did you use pictures, counters, or numbers to help you solve?	1 + 1 = 2	yes	no
Can you explain your thinking in words?		yes	no

Math Vocabulary

counters

Math on the Move Find ways to make 10. Get 10 counters. Separate the counters into 2 piles. You can put 4 counters in one pile and 6 counters in the other. Say, "4 and 6 make 10." Find different ways to make 10.

Add in Any Order

FOLLOWING THE OBJECTIVE
You will add numbers in any order and get the same sum.

LEARN IT: *Addends* are numbers you add. They can be added in any order.

Examples: Use models to add in any order.

Show 4 + 6 = ?	Show 2 + 7 = ?

Show 4 + 6 = ?

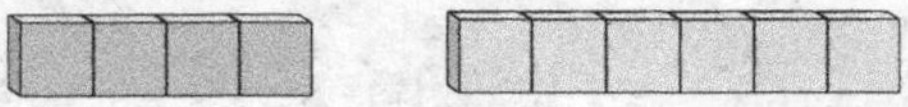

4 + 6 = 10

addends   **sum**

Change the order. 6 + 4 = ?

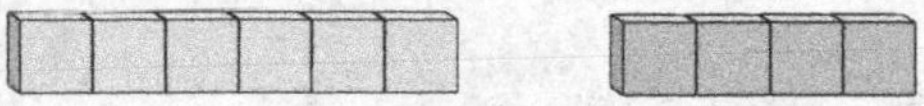

6 + 4 = 10

The sum is the same.

Show 2 + 7 = ?

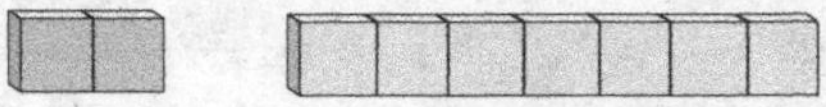

2 + 7 = 9

Change the order of the addends.

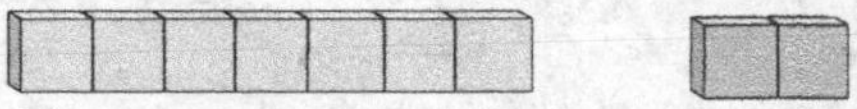

7 + 2 = 9

The sum is the same.

You can change the order of addends. The sum stays the same.

PRACTICE: Now you try

Get two different colored pencils or crayons. Color the blocks to match the problem. Use one color for the first addend. Use a different color for the second addend. Write the sum. Then change the order and add.

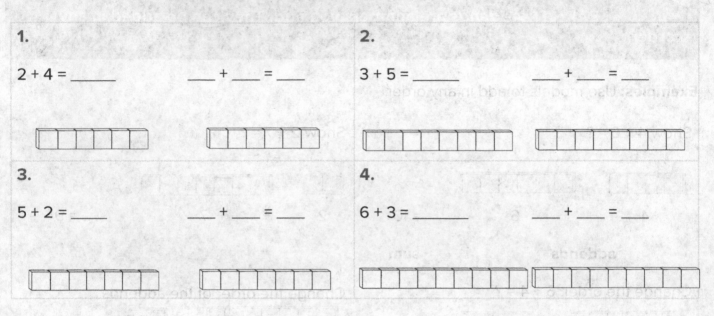

think!
Change the order.
2 + 4 becomes
4 + 2.

1.

2 + 4 = _____   ___ + ___ = ___

2.

3 + 5 = _____   ___ + ___ = ___

3.

5 + 2 = _____   ___ + ___ = ___

4.

6 + 3 = _____   ___ + ___ = ___

Sam and Mark both solved this math problem: *There are 3 dogs at the park. Then 7 more dogs come. How many dogs are at the park now?*

Here are their answers:

Sam's work	Mark's work
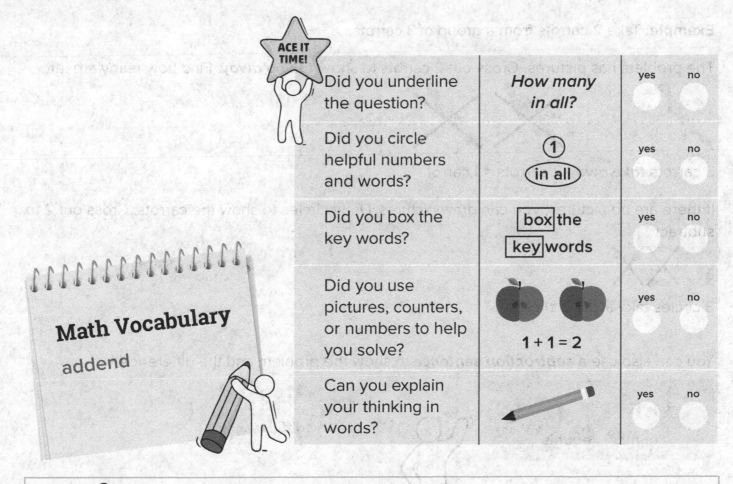	
3 + 7 = 10	7 + 3 = 10

Is Sam's work correct? Is Mark's work correct?

Show your work and explain your thinking here.

Math Vocabulary

addend

			yes	no
Did you underline the question?	*How many in all?*			
Did you circle helpful numbers and words?	① in all		yes	no
Did you box the key words?	box the key words		yes	no
Did you use pictures, counters, or numbers to help you solve?	1 + 1 = 2		yes	no
Can you explain your thinking in words?			yes	no

Math on the Move

Make flashcards with an addition fact like 5 + 3 = 8. Say the "turn around fact": 3 + 5 = 8. This is a good way to memorize addition facts.

Subtraction Concepts

Use Pictures to Subtract

FOLLOWING THE OBJECTIVE
You will use pictures to show subtraction as taking away or taking from.

LEARN IT: You have a group of things. You take some things away from the group. That is *subtraction*. You *subtract* to find the *difference*.

Example: Take 2 carrots from a group of 3 carrots.

The problem has pictures. Cross out 2 carrots to show *taking away*. Find how many are left.

3 carrots *take away* 2 carrots = 1 carrot

If there are no pictures, you can draw pictures. Draw circles to show the carrots. Cross out 2 to subtract.

OXX

3 circles take away 2 circles

> **think!**
> Each circle shows
> 1 carrot.

You can also use a **subtraction sentence** to show the problem and the difference.

3	–	2	=	1
	minus		equals	

> **think!**
> You subtract 2 from 3.
> The difference is 1.

PRACTICE: Now you try

Cross out pictures to subtract. Write the difference.

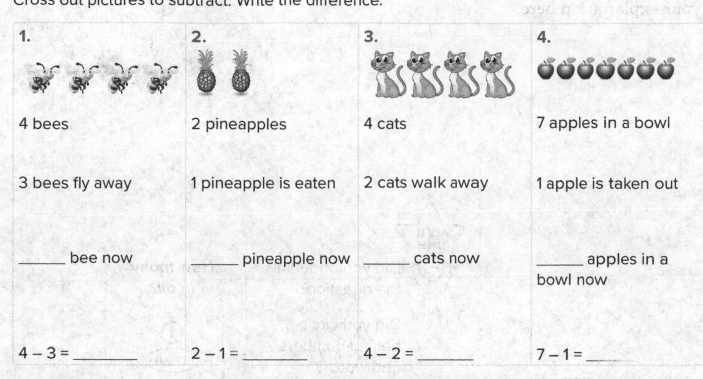

1.

4 bees

3 bees fly away

_____ bee now

4 – 3 = _____

2.

2 pineapples

1 pineapple is eaten

_____ pineapple now

2 – 1 = _____

3.

4 cats

2 cats walk away

_____ cats now

4 – 2 = _____

4.

7 apples in a bowl

1 apple is taken out

_____ apples in a bowl now

7 – 1 = ____

Draw and cross out circles to subtract. Write the difference.

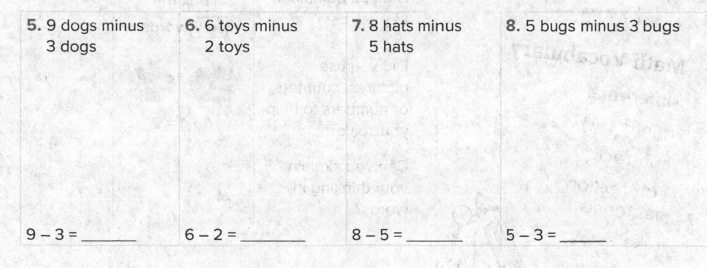

5. 9 dogs minus 3 dogs

9 – 3 = _____

6. 6 toys minus 2 toys

6 – 2 = _____

7. 8 hats minus 5 hats

8 – 5 = _____

8. 5 bugs minus 3 bugs

5 – 3 = _____

There are 9 balloons. Then, 5 balloons pop. How many balloons are left? Show your work and write your explanation here.

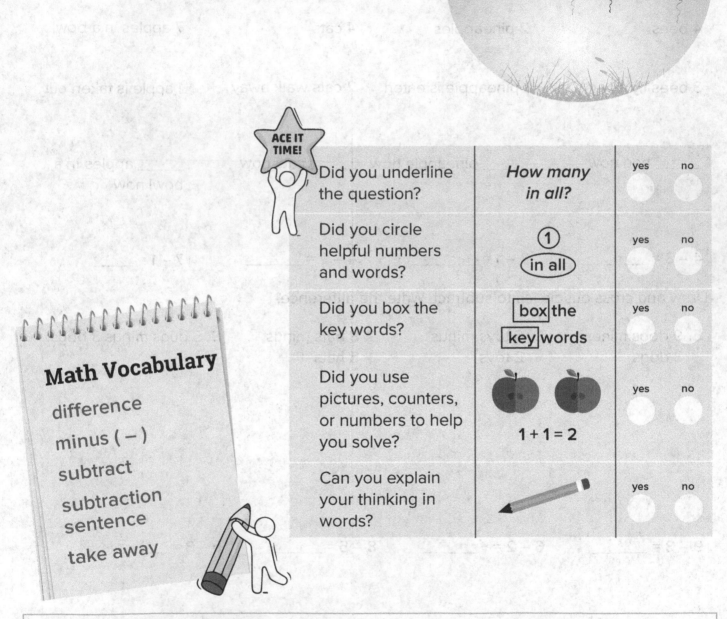

ACE IT TIME!			yes	no
Did you underline the question?	*How many in all?*		○	○
Did you circle helpful numbers and words?	① in all		○	○
Did you box the key words?	box the key words		○	○
Did you use pictures, counters, or numbers to help you solve?	1 + 1 = 2		○	○
Can you explain your thinking in words?			○	○

Math Vocabulary

difference

minus (−)

subtract

subtraction sentence

take away

Math on the Move

You can use real things or people to show subtraction. Ask friends or family members to help you model this problem. There are 5 people in the room. Then, 2 people leave. How many people are left?

Use Counters to Subtract

FOLLOWING THE OBJECTIVE
You will use counters to show subtraction as taking apart.

LEARN IT: You can use counters to *take apart* a group. Place counters in a group. Move some away to show separating the group into 2 parts.

Example: Jenny has 8 plates in all. 2 of her plates are small. The rest are big. How many of Jenny's plates are big?

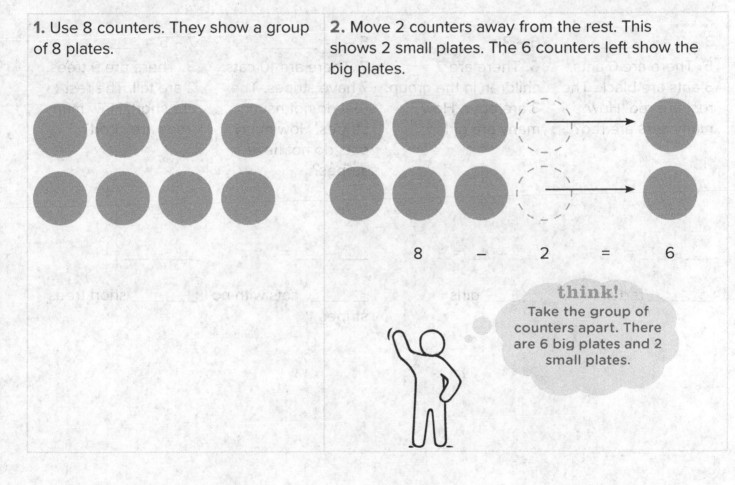

1. Use 8 counters. They show a group of 8 plates.

2. Move 2 counters away from the rest. This shows 2 small plates. The 6 counters left show the big plates.

8 – 2 = 6

think!
Take the group of counters apart. There are 6 big plates and 2 small plates.

PRACTICE: Now you try

Use counters to subtract. Draw and cross out circles to show the counters.

1.	2.	3.	4.
3 – 2 = _____	6 – 3 = _____	5 – 1 = _____	7 – 5 = _____

5. There are 6 ants. 3 ants are black. The rest are red. How many ants are red?

____ – ____ =

_____ red ants

6. There are 7 children in the group. 3 are boys. How many are girls?

____ – ____ =

_____ girls

7. There are 10 cats. 7 have stripes. The rest do not have stripes. How many cats do not have stripes?

____ – ____ =

_____ cats with no stripes

8. There are 9 trees. 2 are tall. The rest are short. How many trees are short?

____ – ____ =

_____ short trees

There are 10 birds in the pond. 2 are ducks. The rest are geese. How many geese are there? Show your work and write your explanation here.

Math Vocabulary

take apart

ACE IT TIME!

	How many in all?	yes	no
Did you underline the question?			
Did you circle helpful numbers and words?	① in all	yes / no	
Did you box the key words?	box the key words	yes / no	
Did you use pictures, counters, or numbers to help you solve?	1 + 1 = 2	yes / no	
Can you explain your thinking in words?		yes / no	

Math on the Move

Get some snacks, like berries or grapes. Make up subtraction problems and act them out. "I have 10 pieces of fruit. 5 are grapes. The rest are berries. How many berries do I have?"

Subtract and Compare

FOLLOWING THE OBJECTIVE
You will compare two numbers or two amounts by subtracting.

LEARN IT: You can use counters to *compare*. You can also draw lines to match. Comparing means deciding how many more or how many less.

Examples:

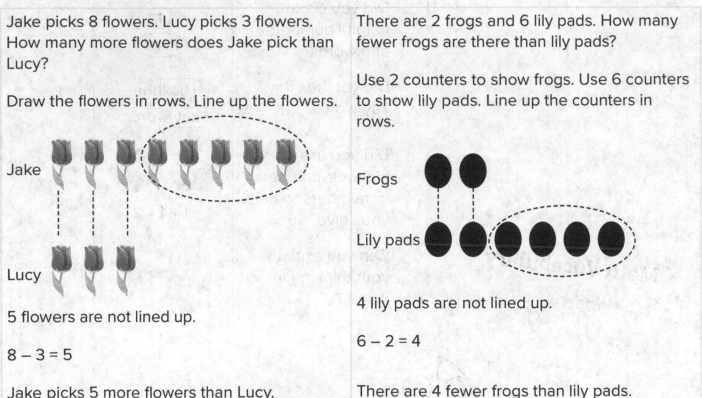

Jake picks 8 flowers. Lucy picks 3 flowers. How many more flowers does Jake pick than Lucy?

Draw the flowers in rows. Line up the flowers.

Jake

Lucy

5 flowers are not lined up.

8 – 3 = 5

Jake picks 5 more flowers than Lucy.

There are 2 frogs and 6 lily pads. How many fewer frogs are there than lily pads?

Use 2 counters to show frogs. Use 6 counters to show lily pads. Line up the counters in rows.

Frogs

Lily pads

4 lily pads are not lined up.

6 – 2 = 4

There are 4 fewer frogs than lily pads.

PRACTICE: Now you try

Subtract. Then compare. You may use counters.

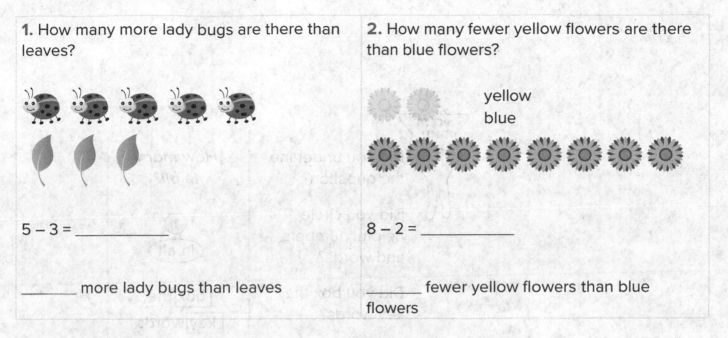

1. How many more lady bugs are there than leaves?

5 − 3 = _____

_____ more lady bugs than leaves

2. How many fewer yellow flowers are there than blue flowers?

yellow

blue

8 − 2 = _____

_____ fewer yellow flowers than blue flowers

Use counters to solve. Write a subtraction sentence to show your work.

3. Amy has 9 pennies and 2 dimes in her bank. How many more pennies does Amy have than dimes?

_____ − _____ = _____

_____ more pennies than dimes

4. Marcus reads 5 books. Rosa reads 7 books. How many fewer books does Marcus read than Rosa?

_____ − _____ = _____

_____ fewer books than Rosa

Mike has 7 stickers. Leah has 10 stickers. How many fewer stickers does Mike have than Leah? Show your work and write your explanation here.

ACE IT TIME!			yes	no
Did you underline the question?	*How many in all?*		yes ○	no ○
Did you circle helpful numbers and words?	① in all		yes ○	no ○
Did you box the key words?	box the key words		yes ○	no ○
Did you use pictures, counters, or numbers to help you solve?	1 + 1 = 2		yes ○	no ○
Can you explain your thinking in words?			yes ○	no ○

Math Vocabulary

compare

Math on the Move

Get 20 counters of any kind. Make up a word problem. For example, "I have 7 crackers and you have 6 crackers. How many more crackers do I have than you?" Use the counters to solve.

Stop and think about what you have learned.

Congratulations! You have finished Units 2 and 3. You can add and subtract. You know how to use drawings and counters to solve problems. You can write addition and subtraction sentences.

Now it's time to show your skills. Solve the problems below! Use what you have learned.

Activity Section 1

1. Write the sum.

2 frogs and 2 more frogs

_____ frogs

2 + 2 = _____

2. Draw a picture to find the sum.

3 + 2 = ____

3. Use counters to add. Draw circles to show your work.

4 + 2 = _____

4. Use counters to solve. Write an addition sentence.

There are 8 red hats and 2 blue hats. How many hats are there in all?

_____ hats

____ + ____ = ____

5. Color the blocks to match the problem.
Use 2 different colors.

4 + 5 = _____

☐☐☐☐☐☐☐☐☐☐

Change the order of the addends. Write the
new problem. Color the blocks to add.

_____ + _____ = _____

☐☐☐☐☐☐☐☐☐☐

6. Solve. Write 2 different addition
sentences to show the problem.

There are 2 dogs. Then 6 dogs come. How
many dogs are there?

_____ dogs

_____ + _____ = _____

_____ + _____ = _____

Activity Section 2

1. Cross out pictures to solve the problem.
Write the difference.

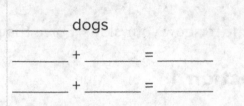

5 apples 1 apple gets eaten

5 – 1 = _____

_____ apples left

2. Draw circles to solve the problem. Write
the difference.

7 bunnies 3 hop away

_____ – _____ = _____

_____ bunnies left

3. Use counters to subtract. Draw circles to
show your work.

9 – 2 = _____

4. Use counters to subtract. There are 10
dogs and 4 are big. The rest are small. How
many small dogs are there?

_____ small dogs

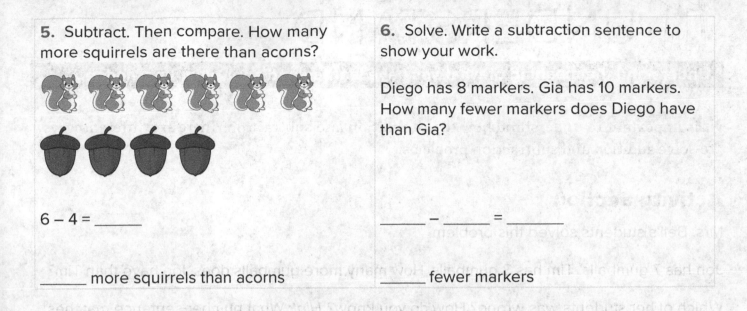

5. Subtract. Then compare. How many more squirrels are there than acorns?

6 – 4 = _____

_____ more squirrels than acorns

6. Solve. Write a subtraction sentence to show your work.

Diego has 8 markers. Gia has 10 markers. How many fewer markers does Diego have than Gia?

_____ – _____ = _____

_____ fewer markers

Activity Section 3

In higher grades, you will add and subtract larger numbers. It will be easier to do this if you already know addition and subtraction facts for the smaller numbers. Practice adding and subtracting numbers up to 10 using counters, pictures, flashcards, or another method. When you are ready, ask an adult to time you while you answer the problems below. How many can you correctly solve in three minutes? Five minutes?

Solve these addition and subtraction facts.

1. 7 + 2 = _____

2. 2 + 3 = _____

3. 10 – 8 = _____

4. 5 + 1 = _____

5. 9 – 5 = _____

6. 3 + 7 = _____

7. 2 + 2 = _____

8. 8 – 3 = _____

9. 7 – 4 = _____

Total Time: _____

UNDERSTAND

Understand the meaning of what you have learned and apply your knowledge.

It is important to understand how to use addition and subtraction. There are different ways to solve addition and subtraction problems.

Activity Section

Mrs. Bell's students solved this problem.

Jon has 7 gumballs. Tim has 3 gumballs. How many more gumballs does Jon have than Tim?

Which of her students was wrong? How do you know? *Hint:* What number sentence matches the problem?

Abby's answer:

$7 - 4 = 3$
I crossed out 3. Jon has 3 more gumballs than Tim.

Nina's answer:

$7 - 3 = 4$
4 gumballs are not lined up. Jon has 4 more gumballs than Tim.

DISCOVER

You can solve problems using addition and subtraction every day. Use what you know to add and subtract.

Activity Section

Tamika has $10. Fill in the table to show what she can buy for $3, $5, or $10. Then write how much change she will get. (*Hint:* The change is $10 minus how much she spent.)

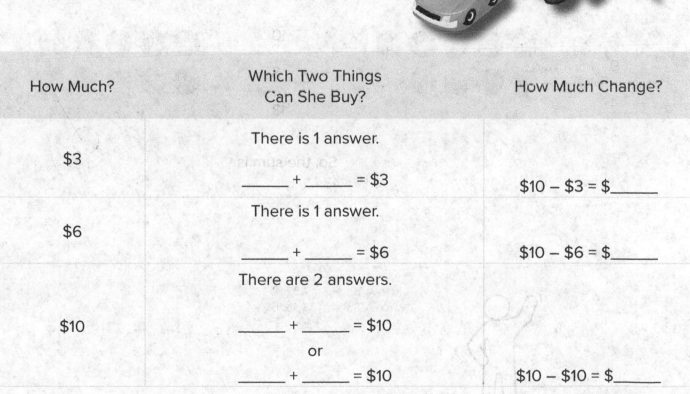

How Much?	Which Two Things Can She Buy?	How Much Change?
$3	There is 1 answer. _____ + _____ = $3	$10 − $3 = $_____
$6	There is 1 answer. _____ + _____ = $6	$10 − $6 = $_____
$10	There are 2 answers. _____ + _____ = $10 or _____ + _____ = $10	$10 − $10 = $_____

Addition and Subtraction Strategies

Use a Double Ten Frame to Add

FOLLOWING THE OBJECTIVE
You will use a double ten frame to add numbers within 20.

LEARN IT: You can use 20 counters and a *double ten frame* to add greater numbers. Use the top *ten frame* to make a 10. Then add.

Example: What is 9 + 4?

1. Put 9 counters in the top ten frame. Then put 4 counters in the bottom ten frame.

$$\begin{array}{r} 9 \\ +\ 4 \\ \hline ? \end{array}$$

2. Move 1 counter to make a 10. Write the new addition problem.

$$\begin{array}{r} 10 \\ +\ 3 \\ \hline 13 \end{array}$$

So, the sum is 13.

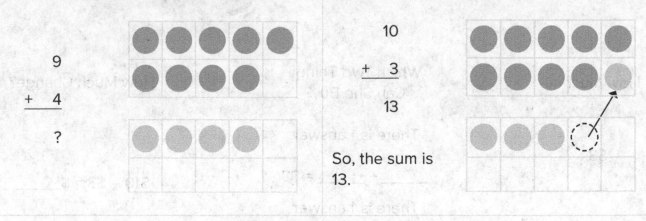

think!
I know 10 + 3 = 13, so I know 9 + 4 = 13, too!

PRACTICE: Now you try

Draw red and yellow circles to model the problem in the first set of double ten frames. Next, move the red counters from the top to the first ten frame at the bottom. Then place yellow counters in the empty spaces to make ten. Place the remaining yellow counters in the bottom ten frame. Write the sum in the boxes.

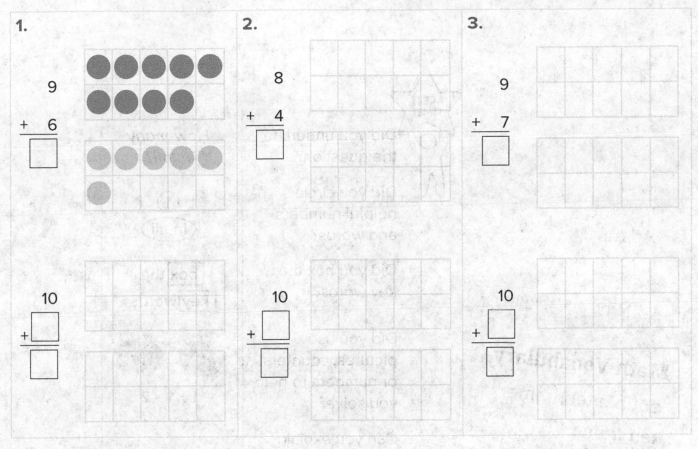

1.

$$\begin{array}{r} 9 \\ +6 \\ \hline \end{array}$$

$$\begin{array}{r} 10 \\ + \\ \hline \end{array}$$

2.

$$\begin{array}{r} 8 \\ +4 \\ \hline \end{array}$$

$$\begin{array}{r} 10 \\ + \\ \hline \end{array}$$

3.

$$\begin{array}{r} 9 \\ +7 \\ \hline \end{array}$$

$$\begin{array}{r} 10 \\ + \\ \hline \end{array}$$

There are 2 big toys and 9 small toys. How many toys are there in all? Use counters to add. Did you put 2 counters or 9 counters in the top frame? Why? Show your work and explain your thinking here.

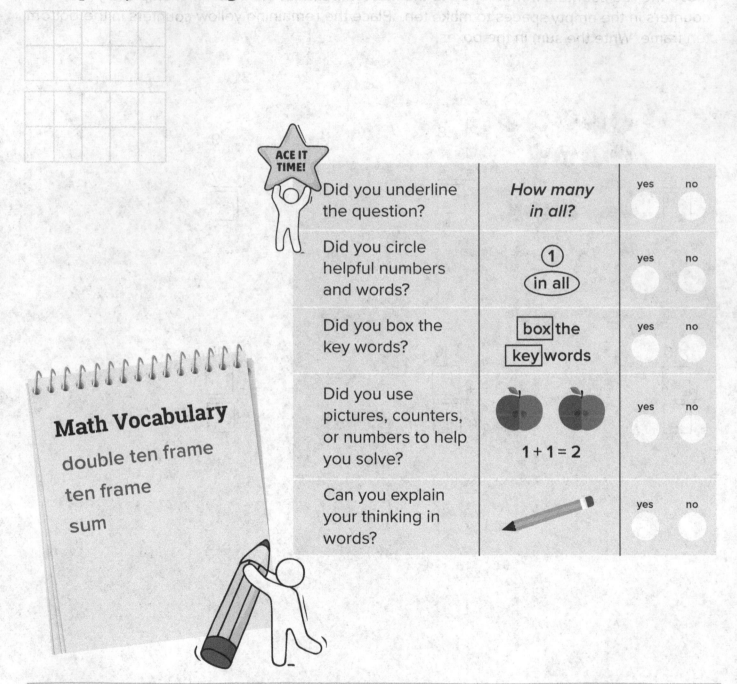

ACE IT TIME!			yes	no
Did you underline the question?	*How many in all?*		yes	no
Did you circle helpful numbers and words?	① in all		yes	no
Did you box the key words?	box the key words		yes	no
Did you use pictures, counters, or numbers to help you solve?	1 + 1 = 2		yes	no
Can you explain your thinking in words?			yes	no

Math Vocabulary

double ten frame

ten frame

sum

Math on the Move

Find two empty 12-egg cartons. Have someone help you cut off 2 cups from each. Now, there are 10 cups in each. Get 20 counters and use the egg cartons as a double ten frame. Solve these problems:
8 + 7 = ?, 8 + 8 = ?, 8 + 9 = ?

Count On to Add

FOLLOWING THE OBJECTIVE
You will add within 20 by counting on from the greater number in an addition sentence.

LEARN IT: You know how to use counters to show two addends. You can also count on from the greater number to find the sum.

Examples:

Add 8 plus 3.	Add 5 plus 7.
⑧ ← Count on from the greater number.	5
+ 3	+ ⑦ ← Count on from the greater number.
Start counting with the number that comes after 8. Count on 3 times.	Count on 5 from 7.
9 comes after 8. So start with 9.	8 comes after 7. Start with 8.
● ● ●	● ● ● ● ●
Count 9, 10, 11	Count 8, 9, 10, 11, 12
8 + 3 = 11	5 + 7 = 12

think!
You can count on using counters, drawings, or your fingers. You can also count on in your head.

PRACTICE: Now you try

Find and circle the greater number. Count on to add.

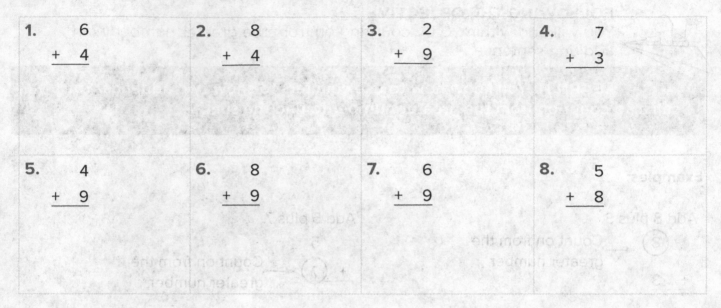

1.	6	2.	8	3.	2	4.	7
	+ 4		+ 4		+ 9		+ 3

5.	4	6.	8	7.	6	8.	5
	+ 9		+ 9		+ 9		+ 8

Marco and Tayo solved the same problem in different ways. Explain why both are correct. Which way do you think is best? Why?

There are 3 boys and 9 girls in the club. How many children are in the club in all?

Marco's way	Tayo's way
Marco counted on from 3. 4, 5, 6, 7, 8, 9, 10, 11, 12 3 + 9 = 12	Tayo counted on from 9. 10, 11, 12 3 + 9 = 12

Show your work and write your explanation here.

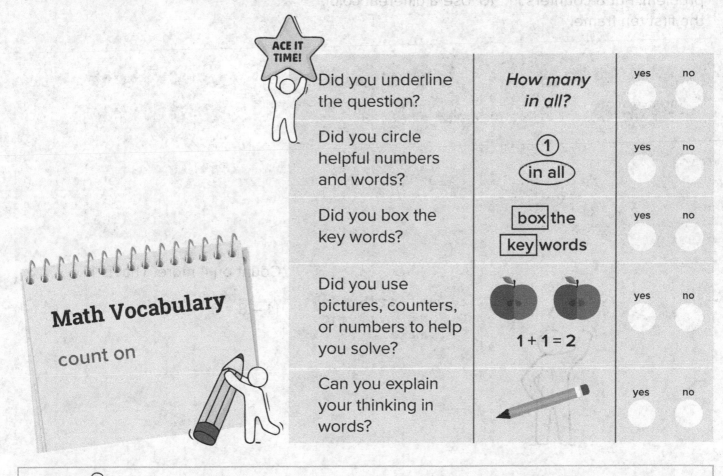

ACE IT TIME!		How many in all?	yes	no
	Did you underline the question?		○	○
	Did you circle helpful numbers and words?	① in all	○	○
	Did you box the key words?	box the key words	○	○
	Did you use pictures, counters, or numbers to help you solve?	1 + 1 = 2	○	○
	Can you explain your thinking in words?		○	○

Math Vocabulary

count on

Math on the Move Practice counting on. Write numbers from 1 to 9 on index cards. Pick two cards at a time. Add them by counting on from the greater number.

Use a Double Ten Frame to Subtract

FOLLOWING THE OBJECTIVE
You will make a 10 to subtract numbers within 20.

LEARN IT: Sometimes, you need to subtract *greater numbers*. You can use 20 counters and a double ten frame. You can make a ten to subtract.

Example: What is 14 − 8?

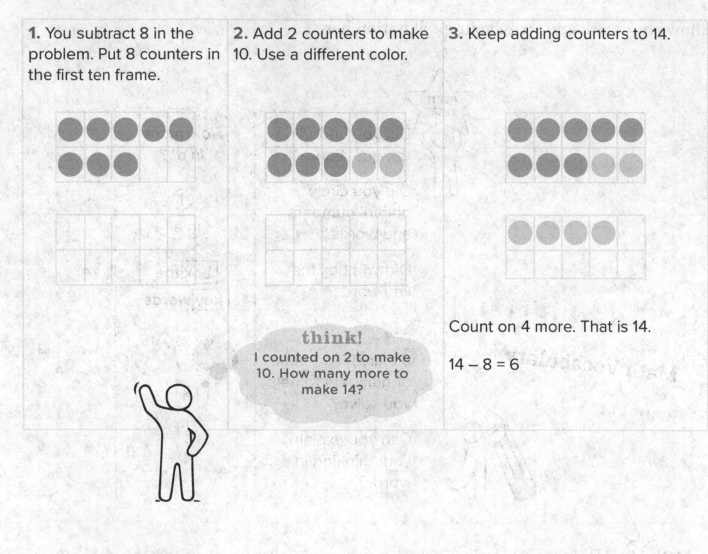

1. You subtract 8 in the problem. Put 8 counters in the first ten frame.

2. Add 2 counters to make 10. Use a different color.

think!
I counted on 2 to make 10. How many more to make 14?

3. Keep adding counters to 14.

Count on 4 more. That is 14.

14 − 8 = 6

PRACTICE: Now you try

Use counters and a double ten frame. Draw red circles to model the number being subtracted. Draw yellow circles to make 10. Then draw more circles to find the difference. Write the difference in the box.

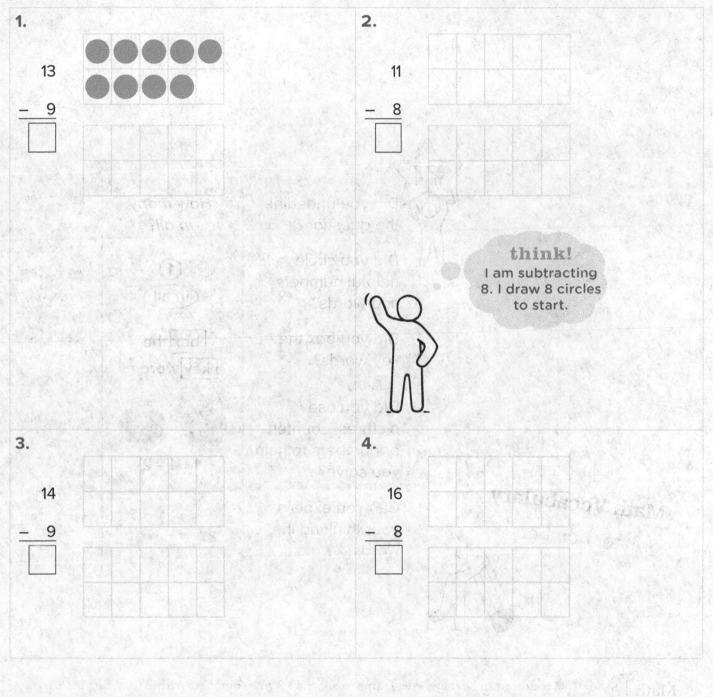

1.

13
− 9

2.

11
− 8

think!
I am subtracting 8. I draw 8 circles to start.

3.

14
− 9

4.

16
− 8

Use a double ten frame to solve the problem. Miguel has 16 beads. 9 are green. The rest are blue. How many beads are blue? Show your work and explain your thinking here.

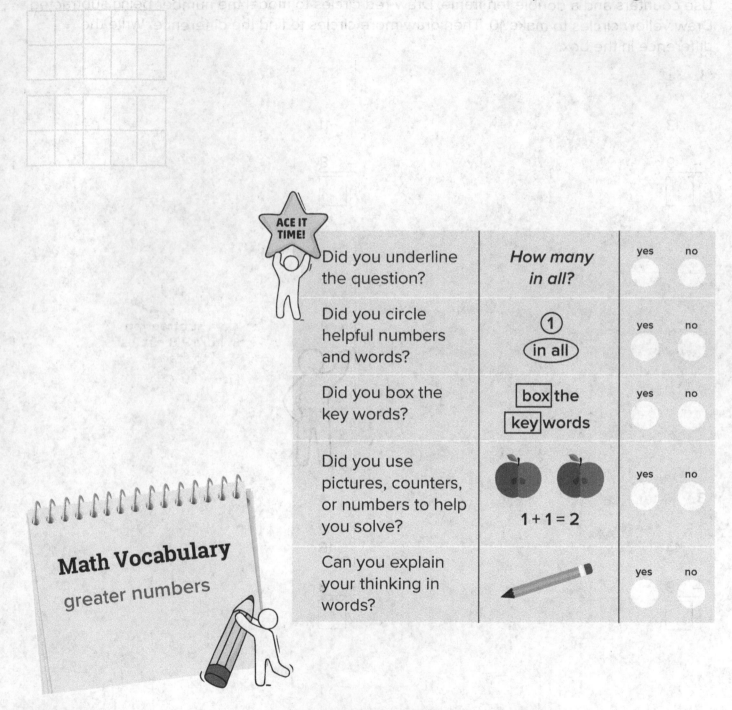

ACE IT TIME!

		yes	no
Did you underline the question?	*How many in all?*	○	○
Did you circle helpful numbers and words?	① in all	○	○
Did you box the key words?	box the key words	○	○
Did you use pictures, counters, or numbers to help you solve?	1 + 1 = 2	○	○
Can you explain your thinking in words?		○	○

Math Vocabulary

greater numbers

Math on the Move

Use counters and a double ten frame to subtract 9 from numbers within 20. Start at 20. Solve 20 – 9 = ?, 19 – 9 = ?, 18 – 9 = ?, and so on. Keep going until you solve 10 – 9 = ? Look at your list of problems. Do you see a pattern?

Use Addition to Subtract

FOLLOWING THE OBJECTIVE
You will use an addition fact to help you subtract.

LEARN IT: Addition can help you subtract. Use an *addition fact* to solve a subtraction problem.

Example: Subtract 12 − 9.

1. Think of a related addition fact.	**2.** Fill in the addition fact and then use it to subtract.
think! Nine plus what equals twelve?	Count on from 9 to get 12.
	10, 11, 12
	You counted on 3 times.
Write the addition fact: 9 + ☐ = 12	Think: 9 + 3 = 12
	Solve: 12 − 9 = 3

PRACTICE: Now you try

Use an addition fact to help you subtract. Write the missing numbers.

1. What is 10 − 8?	**2.** What is 12 − 6?
Think: 8 + ☐ = 10	Think: 6 + ☐ = 12
Solve: 10 − 8 = ☐	Solve: 12 − 6 = ☐

Write an addition fact to help you subtract. Find the difference.

3. What is 11 − 7?	**4.** What is 13 − 5?	**5.** What is 14 − 4?
____ + ____ = ____	____ + ____ = ____	____ + ____ = ____
11 − 7 = _____	13 − 5 = _____	14 − 4 = _____

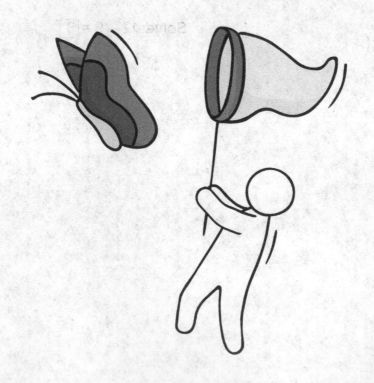

Write an addition fact to solve. There are 14 butterflies. 6 of them are red. The rest are blue. How many butterflies are blue? Show your work and explain your thinking here.

ACE IT TIME!

		yes	no
Did you underline the question?	*How many in all?*	yes	no
Did you circle helpful numbers and words?	① in all	yes	no
Did you box the key words?	box the key words	yes	no
Did you use pictures, counters, or numbers to help you solve?	1 + 1 = 2	yes	no
Can you explain your thinking in words?		yes	no

Math Vocabulary

addition fact

Math on the Move

Make flashcards. Write a subtraction fact on one side. On the other side, write the related addition fact. Practice identifying related facts.

One side	Other side
15 – 6 = _____	6 + _____ = 15

Add Three Numbers

FOLLOWING THE OBJECTIVE
You will add three numbers to find a sum.

LEARN IT: You can add numbers in any *order*. *Group* the numbers to help you add them.

Example: Add 6 + 2 + 4 in two ways.

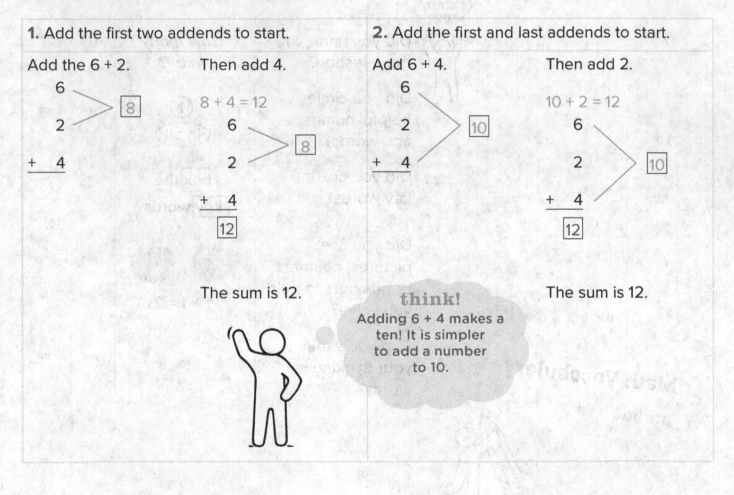

1. Add the first two addends to start.

Add the 6 + 2.

6
2 ⟩ 8
+ 4

Then add 4.

8 + 4 = 12
6
2 ⟩ 8
+ 4
12

The sum is 12.

2. Add the first and last addends to start.

Add 6 + 4.

6
2 ⟩ 10
+ 4

Then add 2.

10 + 2 = 12
6
2 ⟩ 10
+ 4
12

The sum is 12.

think!
Adding 6 + 4 makes a ten! It is simpler to add a number to 10.

PRACTICE: Now you try

Add in two different ways. Add the numbers shown. Write the sum in the box. Then add the other number.

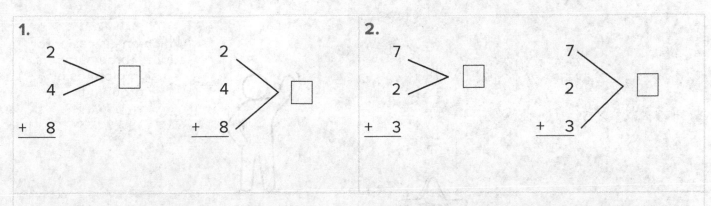

1.
2
4
+ 8

2
4
+ 8

2.
7
2
+ 3

7
2
+ 3

Decide how to add. Draw lines to show which addends you add first. Write the sum in the box. Then add the other number.

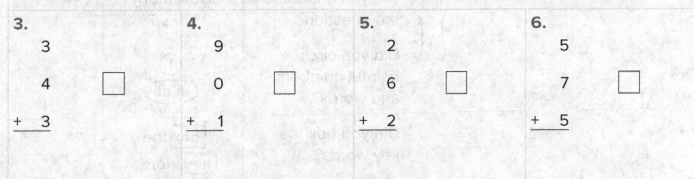

3.
3
4
+ 3

4.
9
0
+ 1

5.
2
6
+ 2

6.
5
7
+ 5

Solve the problem two different ways. First, use pictures to add. Then, use grouping. Mark has 3 red apples, 5 green apples, and 7 yellow apples. How many apples does Mark have in all? Show your work and explain your thinking here.

think!
Which one is easier for adding three numbers, using pictures or grouping?

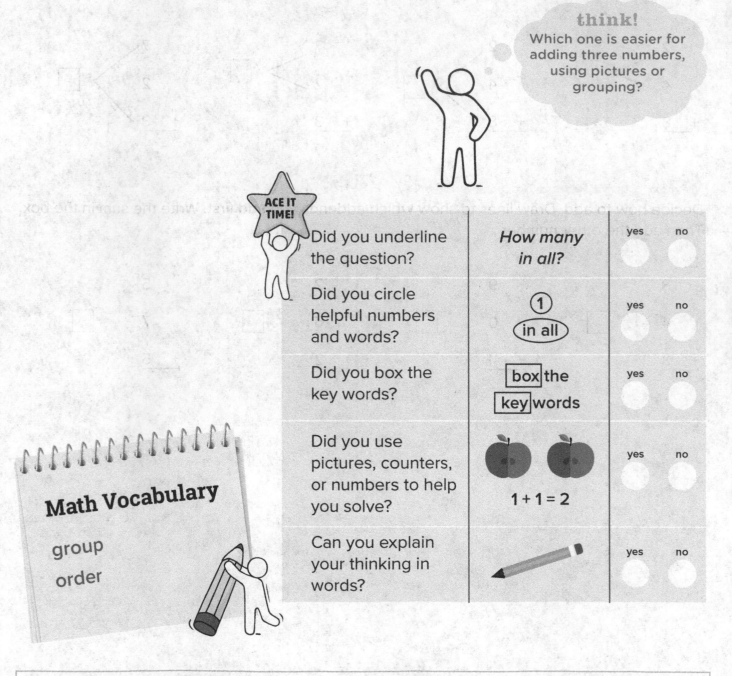

ACE IT TIME!

	How many in all?	yes	no
Did you underline the question?			
Did you circle helpful numbers and words?	① in all	yes	no
Did you box the key words?	box the key words	yes	no
Did you use pictures, counters, or numbers to help you solve?	1 + 1 = 2	yes	no
Can you explain your thinking in words?		yes	no

Math Vocabulary

group

order

Math on the Move   Get a set of number cards. Turn over 3 cards. Add the numbers. Write an addition sentence to show how you added.

Relationships with Operations

UNIT 5

Find Missing Numbers

FOLLOWING THE OBJECTIVE
You will find the missing number in an addition or subtraction sentence.

LEARN IT: Addition and subtraction are related. For example, you can add 5 plus 5 to get 10. You can also subtract 5 from 10 to get 5. So 5 + 5 = 10 and 10 – 5 = 5 are related facts. You can use *related facts* to find missing numbers in problems.

Example:

Marie makes 4 pink bracelets. She also makes some blue bracelets. She makes 11 bracelets in all. How many blue bracelets does she make?

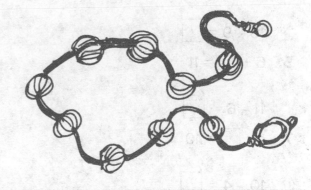

Use an addition fact.	Use a subtraction fact.
Think: There are 4 pink bracelets. She makes some blue bracelets.	Think: There are 11 bracelets.
	4 are pink.
There are 11 bracelets in all.	Some bracelets are blue.
4 + ☐ = 11	11 – 4 = ☐

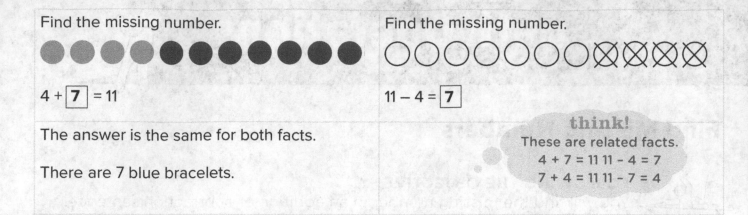

Find the missing number.

$4 + \boxed{7} = 11$

Find the missing number.

$11 - 4 = \boxed{7}$

The answer is the same for both facts.

There are 7 blue bracelets.

think!
These are related facts.
$4 + 7 = 11$ $11 - 4 = 7$
$7 + 4 = 11$ $11 - 7 = 4$

PRACTICE: Now you try

Find the missing numbers. Write the numbers in the boxes.

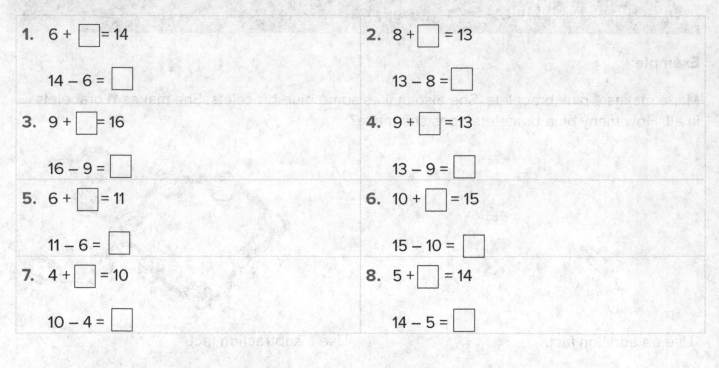

1. $6 + \boxed{} = 14$

   $14 - 6 = \boxed{}$

2. $8 + \boxed{} = 13$

   $13 - 8 = \boxed{}$

3. $9 + \boxed{} = 16$

   $16 - 9 = \boxed{}$

4. $9 + \boxed{} = 13$

   $13 - 9 = \boxed{}$

5. $6 + \boxed{} = 11$

   $11 - 6 = \boxed{}$

6. $10 + \boxed{} = 15$

   $15 - 10 = \boxed{}$

7. $4 + \boxed{} = 10$

   $10 - 4 = \boxed{}$

8. $5 + \boxed{} = 14$

   $14 - 5 = \boxed{}$

Sebastian has 13 marbles. 5 of the marbles are blue. 2 of the marbles are yellow. The rest are gray. How many gray marbles does Sebastian have?

Fill in the blanks in this addition sentence with numbers that you know from the problem. Then solve the problem to find the missing number in the box: Show your work and explain your thinking here.

_____ + _____ + ☐ = _____

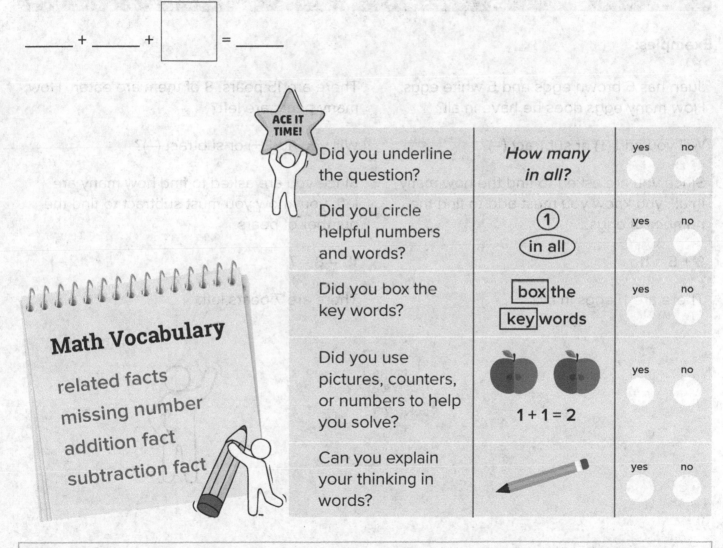

ACE IT TIME!	How many in all?	yes	no
Did you underline the question?	*How many in all?*	○	○
Did you circle helpful numbers and words?	① in all	○	○
Did you box the key words?	box the key words	○	○
Did you use pictures, counters, or numbers to help you solve?	1 + 1 = 2	○	○
Can you explain your thinking in words?		○	○

Math Vocabulary

related facts

missing number

addition fact

subtraction fact

Math on the Move

Find three numbers that make related facts. Write those numbers on one side of an index card. On the other side, write 2 addition facts and 2 subtraction facts. Use the cards to practice the facts.

One side	Other side	
3 7 10	3 + 7 = 10	10 − 3 = 7
	7 + 3 = 10	10 − 7 = 3

Choose an Operation

FOLLOWING THE OBJECTIVE
You will decide to add or subtract to solve word problems.

LEARN IT: Before you solve a *word problem*, you need to determine what question is being asked. Do you need to add or subtract to solve the problem?

Examples:

Juan has 6 brown eggs and 5 white eggs. How many eggs does he have in all?	There are 15 pears. 8 of them are eaten. How many pears are left?
Will you add (+) or subtract (−)?	Will you add (+) or subtract (−)?
Since you are asked to find the how many in all, you know you must add to find the number of eggs.	Since you are asked to find how many are left, you know you must subtract to find the number of pears.
6 + 5 = 11	15 − 8 = 7
There are 11 eggs in all.	There are 7 pears left.

think!
Is there one right answer? You could solve 7 + ☐ = 15, too.

PRACTICE: Now you try

Choose to add or subtract. Write + or − in the circle. Write and solve a number sentence.

1. Manny has 11 crackers. He eats 5 crackers. How many crackers are left?

_____ ◯ _____ = _____

_____ crackers

2. Fay has 8 pencils. Elle has 7 pencils. How many pencils do the girls have in all?

_____ ◯ _____ = _____

_____ pencils

3. Lucas has 14 trucks. He gives 5 of them to his little brother. How many trucks does Lucas have now?

_____ ◯ _____ = _____

_____ trucks

4. Mary has 12 stuffed bears. She buys 4 more bears. How many bears does Mary have now?

_____ ◯ _____ = _____

_____ stuffed bears

Aiden has 8 baseball cards. His brother has 18 cards. How many more cards does Aiden need to have as many as his brother?

Did you add or subtract? Could you have solved it a different way? Show your work and explain your thinking here.

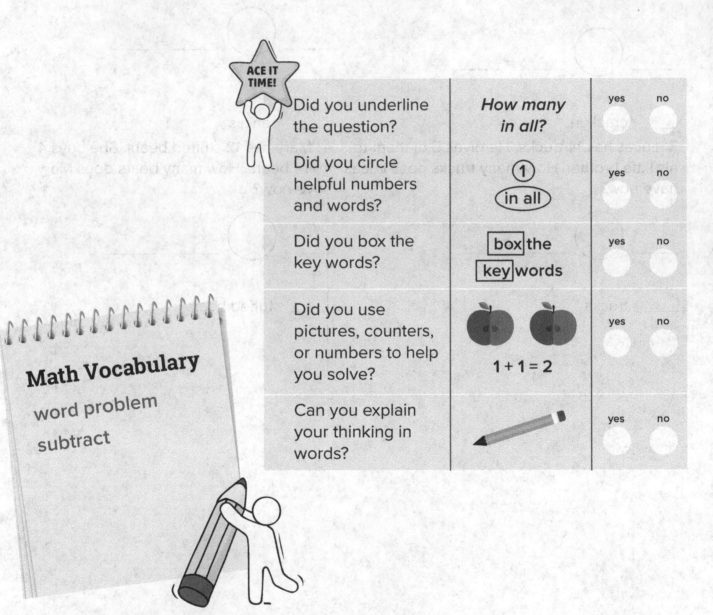

ACE IT TIME!

Did you underline the question?	*How many in all?*	yes	no
Did you circle helpful numbers and words?	① in all	yes	no
Did you box the key words?	box the key words	yes	no
Did you use pictures, counters, or numbers to help you solve?	1 + 1 = 2	yes	no
Can you explain your thinking in words?		yes	no

Math Vocabulary

word problem

subtract

Math on the Move Make up an addition word problem. Then make up a subtraction word problem.

Equal or Not Equal

FOLLOWING THE OBJECTIVE
You will decide if values are equal or not equal. Then you will decide if a number sentence is true or not.

LEARN IT: A *number sentence* has an *equal sign (=).* It shows that both sides are the same. If both sides are the same, the number sentence is *true.* If not, it is *false.* False means not true.

Examples: Decide if each number sentence is true or false.

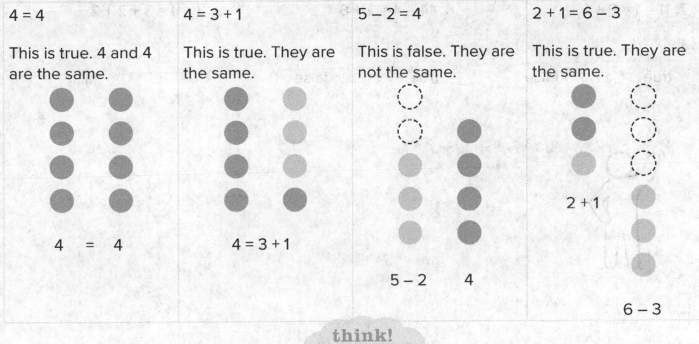

$4 = 4$	$4 = 3 + 1$	$5 - 2 = 4$	$2 + 1 = 6 - 3$
This is true. 4 and 4 are the same.	This is true. They are the same.	This is false. They are not the same.	This is true. They are the same.

$4 \quad = \quad 4$

$4 = 3 + 1$

$5 - 2 \qquad 4$

$2 + 1$

$6 - 3$

think!
3 is not the same as 4.

Unit 5: Relationships with Operations

PRACTICE: Now you try

Circle true or false for each number sentence.

1. $11 = 10$	**2.** $18 - 9 = 9$	**3.** $0 = 7 - 7$
true false	true false	true false
4. $4 + 5 = 5 + 4$	**5.** $10 - 8 = 5 - 2$	**6.** $9 + 6 = 10 + 5$
true false	true false	true false
7. $11 - 1 = 9 + 1$	**8.** $14 - 4 = 2 + 6$	**9.** $3 + 4 = 3 + 2 + 2$
true false	true false	true false

think!
Are both sides equal?

Andrea and Paula count stars in the sky. Andrea counts 5 + 3 + 5.

Paula counts 10 + 3. Did Andrea and Paula count the same number of stars? Explain. Show your work and explain your thinking here.

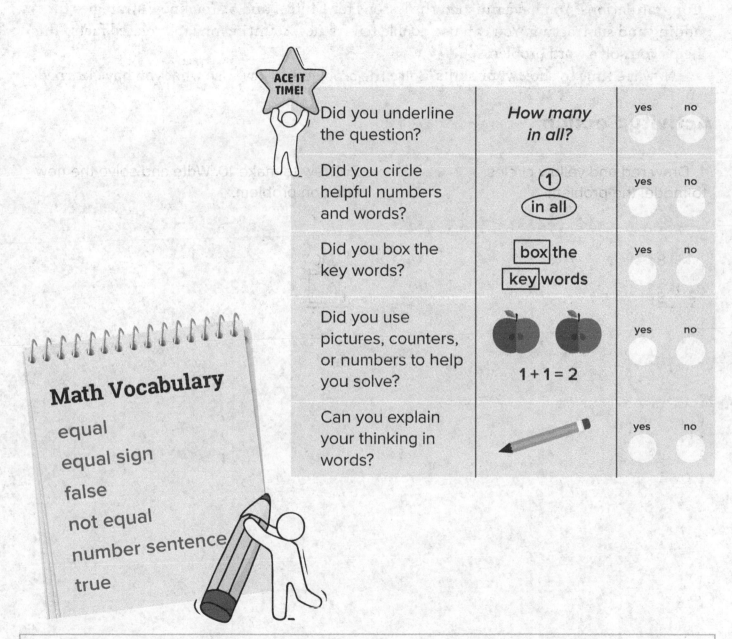

ACE IT TIME!		How many in all?	yes	no
Did you underline the question?			yes	no
Did you circle helpful numbers and words?		① in all	yes	no
Did you box the key words?		box the key words	yes	no
Did you use pictures, counters, or numbers to help you solve?		1 + 1 = 2	yes	no
Can you explain your thinking in words?			yes	no

Math Vocabulary

equal

equal sign

false

not equal

number sentence

true

Math on the Move

Get a piece of paper and write true number sentences that have sums or differences of 10. Then do this for every number from 11 to 15.

5 + 5 = 10	3 + 7 = 10
6 + 4 = 10	2 + 8 = 10
15 − 5 = 10	11 − 1 = 10

REVIEW

Congratulations! You have finished the lessons for Units 4 and 5. You know strategies for adding and subtracting. You can use double ten frames, counting on, and related facts. This helps you solve word problems.

Now it's time to show your skills. Solve the problems below! Use what you have learned.

Activity Section

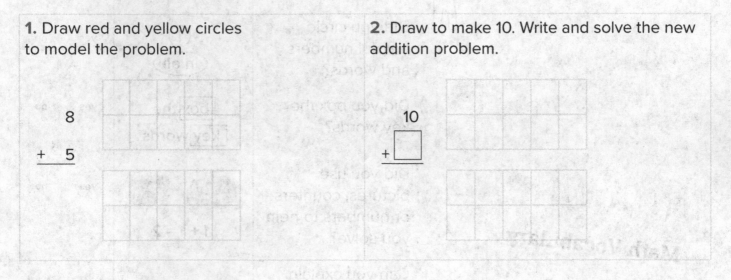

1. Draw red and yellow circles to model the problem.

```
   8
+  5
────
```

2. Draw to make 10. Write and solve the new addition problem.

```
  10
+ □
```

3. Draw red and yellow circles to model the problem. Draw yellow circles to make 10. Find the difference.

12

– 9

4. Write an addition fact to help you subtract. Find the difference.

What is 11 – 5?

_____ + _____ = _____

11 – 5 = _____

Find and circle the greater number. Count on to add.

5.

7

+ 4

6.

5

+ 9

Draw lines to show which two addends you add first. Find the total sum.

7.

6

3

+ 3

8.

8

4

+ 6

Find the missing number.

9.

$5 + \boxed{} = 12$

$12 - 5 = \boxed{}$

10.

$14 = 6 + \boxed{}$

$\boxed{} = 14 - 6$

Write + or − in the circle. Write and solve a number sentence.

11. Brian sees 9 deer. Rob sees 6 deer. How many deer do the boys see all together?

_____ ◯ _____ = _____

_____ deer

12. Elena has 19 fish. 9 of them are blue. The rest are orange. How many orange fish are there?

_____ ◯ _____ = _____

_____ orange fish

Circle true or false.

13. $9 + 4 = 10 + 4$

   true   false

14. $7 + 7 + 5 = 14 + 7$

   true   false

15. $10 - 3 = 9 - 4$

   true   false

16. $11 - 1 - 3 = 7$

   true   false

UNDERSTAND

Understand the meaning of what you have learned and apply your knowledge.

It is important to understand how to use addition and subtraction to solve problems. There are different ways to do this.

Activity Section

Mei solves this problem.

Rashad invites 9 boys to his party. Then he invites some girls. He invites 15 children in all. How many girls does he invite?

Mei's answer:

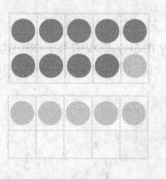

15 − 9 = ?

I put 9 counters in the ten frame. I added 1 more to make 10. I added 5 more to make 15. There are 6 more counters. 15 − 9 = 6. He invites 6 girls.

Mei subtracted to solve the problem. Mei says this is the only way to solve the problem. Is that true? If not, show another way. Write your answer and explain your thinking here.

DISCOVER

We solve problems using addition and subtraction every day. Use the strategies you have learned to add and subtract

Activity Section

There are 15 children at lunch. 7 children eat pizza. The rest eat sandwiches or soup.

How many children eat sandwiches or soup? Write and solve a number sentence.

_____ _____ = _____

_____ children eat sandwiches or soup.

How many children could be eating sandwiches? How many could be eating soup? Write a possible answer below. (There is more than one answer.) Show your work and write your explanation here.

| 7 children eat pizza. | _____ children eat sandwiches. | _____ children eat soup. |

Write an addition sentence to show that your answer equals 15.

_____ + _____ + _____ = 15

Count by Ones and Tens to 120

FOLLOWING THE OBJECTIVE
You will count by ones or tens to 120.

LEARN IT: You can count by *ones* or *tens* starting at any number. Use the 120 chart to help you.

1	2	3	4	5	6	7	8	9	10
11	12	13	14	15	16	17	18	19	20
21	22	23	24	25	26	27	28	29	30
31	32	33	34	35	36	37	38	39	40
41	42	43	44	45	46	47	48	49	50
51	52	53	54	55	56	57	58	59	60
61	62	63	64	65	66	67	68	69	70
71	72	73	74	75	76	77	78	79	80
81	82	83	84	85	86	87	88	89	90
91	92	93	94	95	96	97	98	99	100
101	102	103	104	105	106	107	108	109	110
111	112	113	114	115	116	117	118	119	120

Example: Fill in the blanks using the 120 chart:

63, 64, 65, ___, ___, ___, ___

1. Find the numbers in the 120 chart.

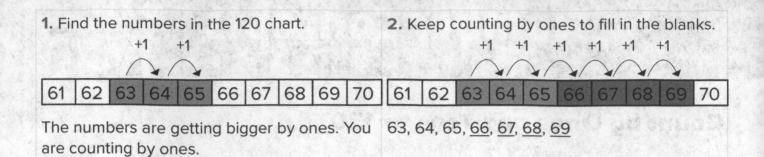

| 61 | 62 | 63 | 64 | 65 | 66 | 67 | 68 | 69 | 70 |

The numbers are getting bigger by ones. You are counting by ones.

2. Keep counting by ones to fill in the blanks.

| 61 | 62 | 63 | 64 | 65 | 66 | 67 | 68 | 69 | 70 |

63, 64, 65, <u>66</u>, <u>67</u>, <u>68</u>, <u>69</u>

Example: Fill in the blanks using the 120 chart:

30, 40, 50, ___, ___, ___, ___

1. Find the numbers in the 120 chart.

| 20 |
| 30 | +10 |
| 40 | +10 |
| 50 |
| 60 |
| 70 |
| 80 |
| 90 |
| 100 |

The numbers are getting bigger by tens. You are counting by tens.

2. Keep counting by tens to fill in the blanks.

| 20 |
30	+10
40	+10
50	+10
60	+10
70	+10
80	+10
90	+10
100

30, 40, 50, <u>60</u>, <u>70</u>, <u>80</u>, <u>90</u>

PRACTICE: Now you try

Use a 120 chart. Count by ones or tens to fill in the blanks.

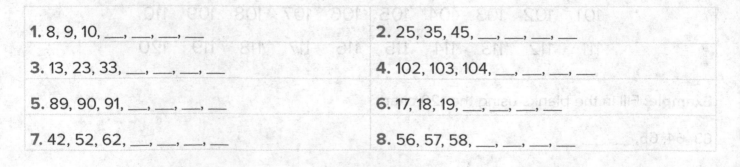

1. 8, 9, 10, ___, ___, ___, ___

2. 25, 35, 45, ___, ___, ___, ___

3. 13, 23, 33, ___, ___, ___, ___

4. 102, 103, 104, ___, ___, ___, ___

5. 89, 90, 91, ___, ___, ___, ___

6. 17, 18, 19, ___, ___, ___, ___

7. 42, 52, 62, ___, ___, ___, ___

8. 56, 57, 58, ___, ___, ___, ___

There are 10 coins in each bag. You can find the total coins by counting by tens. How many coins are in the bags in all? Show your work and write your explanation here.

_____ coins

ACE IT TIME!		How many in all?	yes	no
	Did you underline the question?		◯	◯
	Did you circle helpful numbers and words?	① (in all)	◯	◯
	Did you box the key words?	box the key words	◯	◯
	Did you use pictures, counters, or numbers to help you solve?	1 + 1 = 2	◯	◯
	Can you explain your thinking in words?		◯	◯

Math Vocabulary

count

ones

tens

Math on the Move

Pretend you are planning a party for your class. You need one slice of pizza for each student. Count the number of students in your class to find the number of slices you need. Do you count by ones or count by tens?

Tens and Ones Through 120

FOLLOWING THE OBJECTIVE
You will understand that the digits in two-digit numbers show the amounts of tens and ones.

LEARN IT: You can break a two-digit number into tens and ones. The *digits* tell you the amount of tens or ones.

Example: How many tens and ones are in the number 48?

1. Use a picture to show 48.	**2.** Count the tens. There are 4 tens.
3. Count the ones. There are 8 ones.	**4.** Look at the number. first digit second digit **48** The first digit shows 4 tens. The second digit shows 8 ones. Another way to write 48 is 4 tens 8 ones.

PRACTICE: Now you try

Find the amount of tens and ones in each number.

1. 17 __ tens __ ones	**2.** 83 __ tens __ ones
3. 55 __ tens __ ones	**4.** 32 __ tens __ ones
5. 79 __ tens __ ones	**6.** 90 __ tens __ ones
7. 64 __ tens __ ones	**8.** 11 __ ten __ one

A number has 2 tens and 6 ones. What is the number? Show your work and write your explanation here.

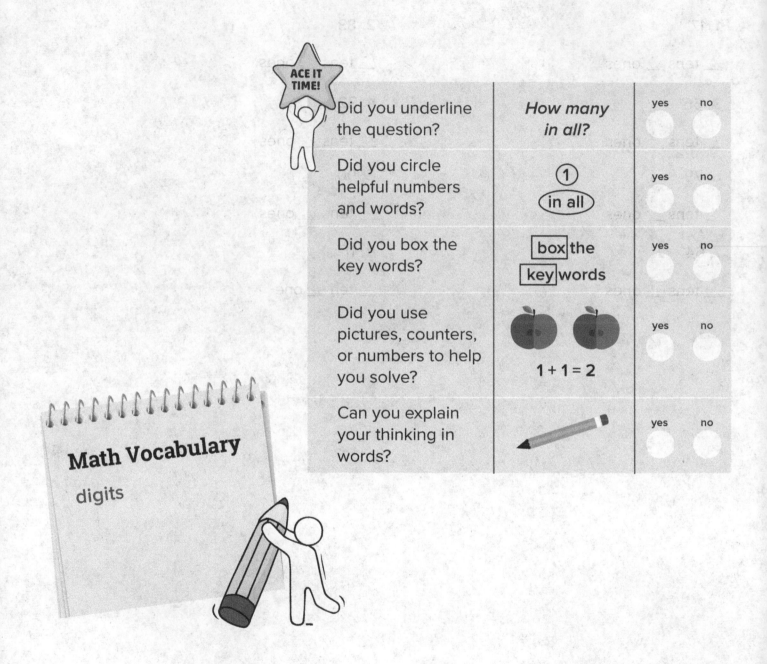

		yes	no
Did you underline the question?	*How many in all?*	○	○
Did you circle helpful numbers and words?	① in all	○	○
Did you box the key words?	box the key words	○	○
Did you use pictures, counters, or numbers to help you solve?	1 + 1 = 2	○	○
Can you explain your thinking in words?		○	○

Math Vocabulary

digits

Math on the Move

You can use coins to understand tens and ones. Pennies are each 1 cent, and dimes are each 10 cents. Use dimes and pennies to create different two-digit numbers. Compare the amount of tens and ones to the number of dimes and pennies.

Compare Numbers Using >, <, =

FOLLOWING THE OBJECTIVE
You will compare two-digit numbers using >, <, and =.

LEARN IT: You can compare numbers. Use the symbols **> (greater than)**, **< (less than)**, and **= (equal to)**. Use the amounts of tens and ones to compare two-digit numbers.

Examples: Compare the numbers. Write >, <, or = in each circle.

9 ◯ 5	5 ◯ 9	9 ◯ 9
9 is greater than 5.	5 is less than 9.	9 is equal to 9.
So, 9 > 5.	So, 5 < 9.	So, 9 = 9.

Example: Write >, <, or = in the circle.

36 ◯ 39.

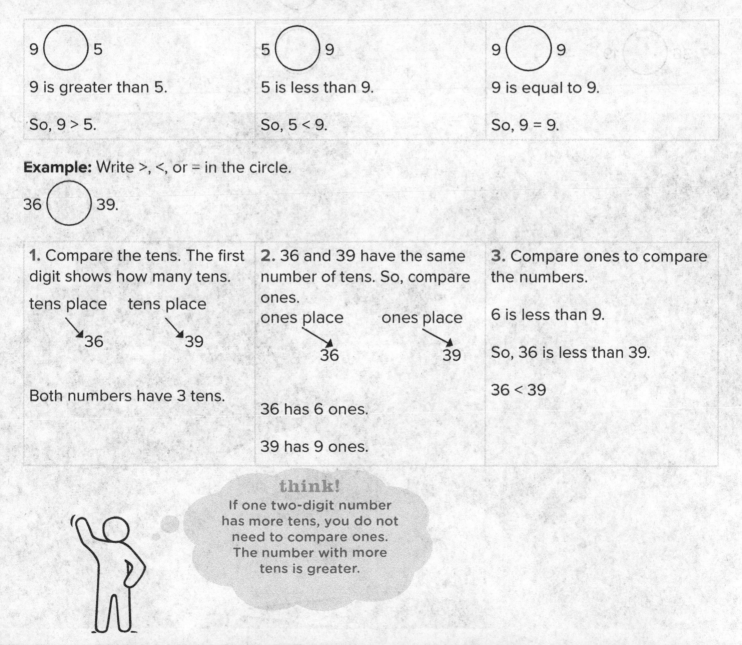

1. Compare the tens. The first digit shows how many tens.

tens place   tens place

↘36   ↘39

Both numbers have 3 tens.

2. 36 and 39 have the same number of tens. So, compare ones.

ones place   ones place

↘36   ↘39

36 has 6 ones.

39 has 9 ones.

3. Compare ones to compare the numbers.

6 is less than 9.

So, 36 is less than 39.

36 < 39

think!
If one two-digit number has more tens, you do not need to compare ones. The number with more tens is greater.

PRACTICE: Now you try

Compare the two numbers. Write >, <, or = in the circle.

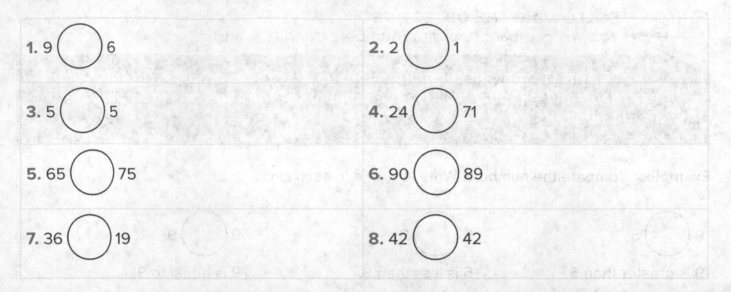

1. 9 ◯ 6

2. 2 ◯ 1

3. 5 ◯ 5

4. 24 ◯ 71

5. 65 ◯ 75

6. 90 ◯ 89

7. 36 ◯ 19

8. 42 ◯ 42

Carlos got 43 correct answers on a test. Luke got 39 correct answers on the same test. Who got more correct answers, Carlos or Luke? Compare the numbers using >, <, and =. Show your work and write your explanation here.

ACE IT TIME!

		yes	no
Did you underline the question?	*How many in all?*	◯	◯
Did you circle helpful numbers and words?	① in all	◯	◯
Did you box the key words?	box the key words	◯	◯
Did you use pictures, counters, or numbers to help you solve?	$1 + 1 = 2$	◯	◯
Can you explain your thinking in words?		◯	◯

Math Vocabulary

greater than

less than

equal to

two-digit numbers

Math on the Move

Think of some of your family members' ages. Write them down. Then use >, <, and = to compare their ages. Find out who is the oldest and who is the youngest.

Two-Digit Addition and Subtraction

Add and Subtract Tens

FOLLOWING THE OBJECTIVE
You will add and subtract two-digit numbers made with tens.

LEARN IT: You can use models to add and subtract tens.

Example: Use a model to find 80 – 20.

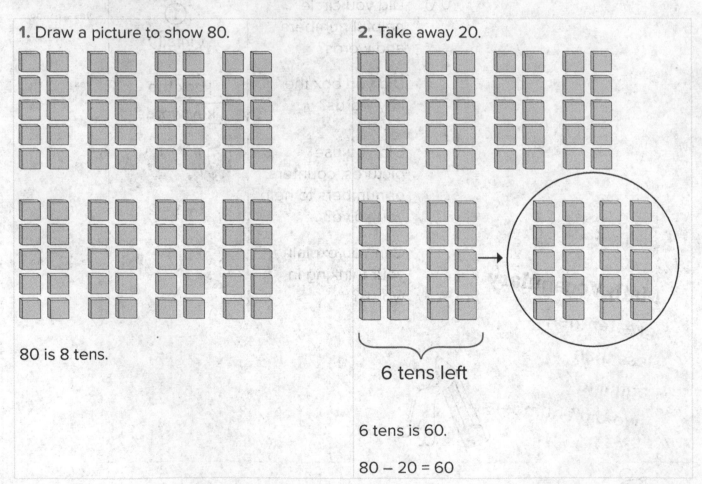

1. Draw a picture to show 80.

80 is 8 tens.

2. Take away 20.

6 tens left

6 tens is 60.

80 – 20 = 60

PRACTICE: Now you try

Draw a picture to add or subtract. Use a separate piece of paper.

1. 10 + 70 = _____

2. 30 + 40 = _____

3. 50 – 50 = _____

4. 20 – 10 = _____

5. 90 – 80 = _____

6. 60 + 20 = _____

7. 40 – 20 = _____

8. 30 + 30 = _____

Dave does 40 jumping jacks. Liz does 20 more jumping jacks than Dave. How many jumping jacks does Liz do? Show your work and write your explanation here.

_____ jumping jacks

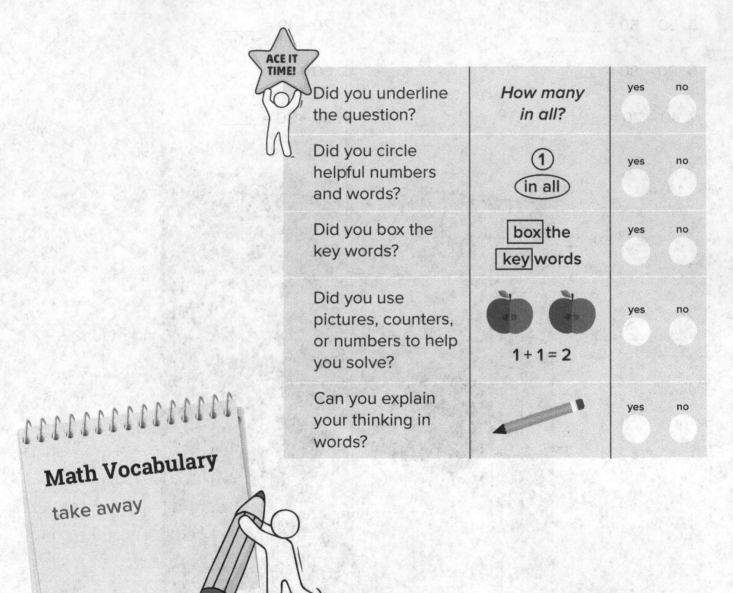

ACE IT TIME!		How many in all?	yes	no
	Did you underline the question?		○	○
	Did you circle helpful numbers and words?	① in all	○	○
	Did you box the key words?	box the key words	○	○
	Did you use pictures, counters, or numbers to help you solve?	1 + 1 = 2	○	○
	Can you explain your thinking in words?		○	○

Math Vocabulary

take away

Math on the Move Dimes are each 10 cents. Get 9 dimes from an adult. Use the dimes to show subtraction of tens. How many dimes are left after you take away 2? How many cents is that?

Mental Math: Add 10 or Subtract 10

FOLLOWING THE OBJECTIVE
You will add or subtract 10 without counting.

LEARN IT: You can use *mental math* to add or subtract 10 from a two-digit number. Mental math means that you do not draw or write on paper. You solve the problem in your head.

Examples:

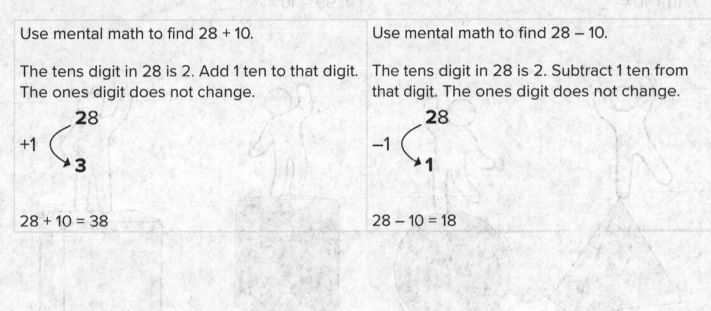

Use mental math to find 28 + 10.

The tens digit in 28 is 2. Add 1 ten to that digit. The ones digit does not change.

28
+1
3

28 + 10 = 38

Use mental math to find 28 − 10.

The tens digit in 28 is 2. Subtract 1 ten from that digit. The ones digit does not change.

28
−1
1

28 − 10 = 18

PRACTICE: Now you try

Use mental math to add and subtract.

1. 35 + 10 = _____

2. 51 – 10 = _____

3. 79 + 10 = _____

4. 64 – 10 = _____

5. 48 – 10 = _____

6. 82 + 10 = _____

7. 10 + 10 = _____

8. 99 – 10 = _____

43 students are on a bus. The bus driver lets 10 students off the bus. How many students are left on the bus now? Show your work and write your explanation here.

_____ students

ACE IT TIME!			yes	no
Did you underline the question?	*How many in all?*		yes ○	no ○
Did you circle helpful numbers and words?	① in all		yes ○	no ○
Did you box the key words?	box the key words		yes ○	no ○
Did you use pictures, counters, or numbers to help you solve?	🍎 🍎 1 + 1 = 2		yes ○	no ○
Can you explain your thinking in words?	✏️		yes ○	no ○

Math Vocabulary

mental math

Math on the Move

Collect pencils in your classroom. Make sure you have between 10 and 20. Ask a friend to take away 10 pencils. Can you figure out how many pencils are left without counting?

Use Tens and Ones to Add

FOLLOWING THE OBJECTIVE
You will add two-digit numbers using tens and ones.

LEARN IT: The place of a digit tells you its value. This is called *place value*. You can use place value to add numbers. Tens are added to other tens. Ones are added to other ones.

Example: Use place value to add 48 + 15.

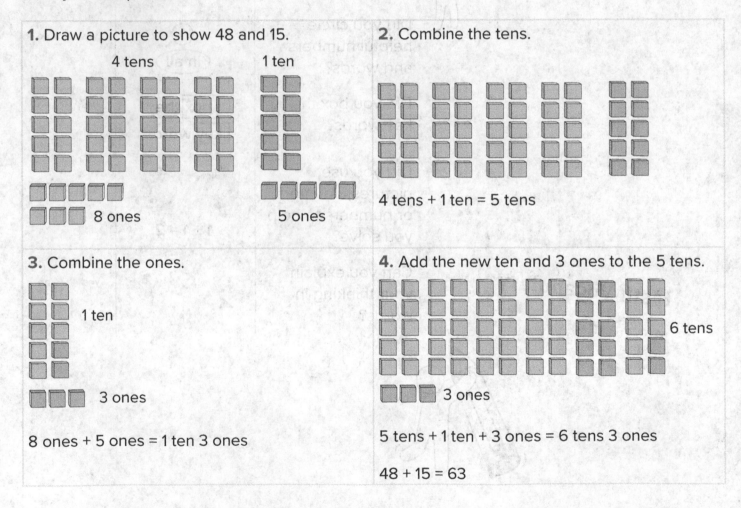

1. Draw a picture to show 48 and 15.

4 tens   1 ten

8 ones   5 ones

2. Combine the tens.

4 tens + 1 ten = 5 tens

3. Combine the ones.

1 ten

3 ones

8 ones + 5 ones = 1 ten 3 ones

4. Add the new ten and 3 ones to the 5 tens.

6 tens

3 ones

5 tens + 1 ten + 3 ones = 6 tens 3 ones

48 + 15 = 63

PRACTICE: Now you try

Use the amount of tens and ones to add the numbers.

1. 3 + 12 = _____   **2.** 25 + 11 = _____

3. 46 + 7 = _____   **4.** 34 + 34 = _____

5. 29 + 51 = _____   **6.** 65 + 17 = _____

7. 73 + 9 = _____   **8.** 42 + 24 = _____

Peter and Olivia are playing a game together. Peter earns 35 points. Olivia earns 41 points. How many points did Peter and Olivia earn in all? Show your work and write your explanation here.

_____ points

ACE IT TIME!		yes	no
Did you underline the question?	*How many in all?*	○	○
Did you circle helpful numbers and words?	①in all	○	○
Did you box the key words?	box the key words	○	○
Did you use pictures, counters, or numbers to help you solve?	1 + 1 = 2	○	○
Can you explain your thinking in words?		○	○

Math Vocabulary

place value

Math on the Move

Count the students in your classroom. Count the students in another classroom in first grade. How many students are in the two classes in all?

REVIEW

Congratulations! You've finished the lessons for this unit. This means you've learned about counting by ones and tens. You've used digits in a number to find the amount of tens and ones in that number. You've used pictures to compare two-digit numbers, and to help you add and subtract tens. You've used mental math to add or subtract 10, and used place value to add two-digit numbers.

Now it's time to prove your skills with two-digit numbers. Solve the problems below using everything you have learned.

Activity Section 1

Count by ones to find the next numbers.

1. 113, 114, 115, ____, ____, ____

2. 38, 39, 40, ____, ____, ____

3. 10, 11, 12, ____, ____, ____

4. 85, 86, 87, ____, ____, ____

Count by tens to find the next numbers.

5. 8, 18, 28, ____, ____, ____

6. 26, 36, 46, ____, ____, ____

7. 55, 65, 75, ____, ____, ____

8. 30, 40, 50, ____, ____, ____

Activity Section 2

Find the amount of tens and ones in each number.

1. 80 ___ tens ___ ones	**2.** 29 ___ tens ___ ones
3. 37 ___ tens ___ ones	**4.** 93 ___ tens ___ ones
5. 12 ___ tens ___ ones	**6.** 72 ___ tens ___ ones
7. 45 ___ tens ___ ones	**8.** 68 ___ tens ___ ones

Activity Section 3

Fill in the circle with >, <, or = .

1. 30 ◯ 19	**2.** 23 ◯ 23
3. 58 ◯ 91	**4.** 84 ◯ 85
5. 62 ◯ 62	**6.** 47 ◯ 74
7. 14 ◯ 20	**8.** 78 ◯ 72

Activity Section 4

Add or subtract.

1. 50 – 30 = _____	**2.** 10 + 30 = _____
3. 20 + 40 = _____	**4.** 70 – 60 = _____
5. 40 – 40 = _____	**6.** 80 – 50 = _____
7. 20 + 20 = _____	**8.** 90 – 10 = _____

Activity Section 5

Add or subtract 10 using mental math.

1. 41 + 10 = _____	**2.** 36 – 10 = _____
3. 85 – 10 = _____	**4.** 53 + 10 = _____
5. 22 – 10 = _____	**6.** 19 + 10 = _____
7. 68 + 10 = _____	**8.** 79 – 10 = _____

Activity Section 6

Add.

1. 22 + 39 = _____	**2.** 8 + 90 = _____
3. 17 + 16 = _____	**4.** 63 + 27 = _____
5. 45 + 45 = _____	**6.** 52 + 29 = _____
7. 86 + 11 = _____	**8.** 48 + 36 = _____

UNDERSTAND

Understand the meaning of what you have learned and apply your knowledge.

Activity Section

Cameron saved some money. He saved $10 bills and $1 bills. The money he saved is in the picture. How much money did Cameron save?

Cameron spends $20 on a new game. He now has $20 less. How much money does he have now?

Show your work and write your explanation in the space below.

DISCOVER

Activity Section

Two classes are going on a field trip: Mr. Won's class and Mrs. Smith's class. There are 27 students in Mr. Won's class. There are 24 students in Mrs. Smith's class. How many students are there in all?

The classes want to take a bus that holds 56 students. Can all the students fit on the bus?

Show your work and explain your thinking in the space provided.

Measurement and Data Concepts

Order and Compare Lengths

FOLLOWING THE OBJECTIVE
You will order and compare the lengths of three objects.

LEARN IT: One way you can *measure* objects is to compare their *lengths*.

Example: These three line segments are *ordered* from *shortest* to *longest*. Notice how they all start from the same place.

Shortest	———————————
	—————————————————————————
Longest	———————————————————————————————

But what if we are only given two lengths to compare? How can we find the length of a third object?

Example: Draw a green line that is longer than the red line, and shorter than the blue line.

Red	————————————
Green	– – – – – – – – – – –
Blue	———————————————————————

Trace the dotted green line to show the possible length of the green line.

PRACTICE: Now you try

1. Look at the crayons. Fill in the blanks.

blue

yellow

red

The _____ crayon is the shortest.

The _____ crayon is the longest.

2. Draw three lines in order from **shortest** to **longest**.

Shortest	
Longest	

3. Draw three lines in order from **longest** to **shortest**.

Longest	
Shortest	

A green pencil is shorter than a blue pencil. An orange pencil is shorter than the green pencil. Which is the longest pencil? How do you know? Draw lines to help you determine the answer. Show your work and write your explanation here.

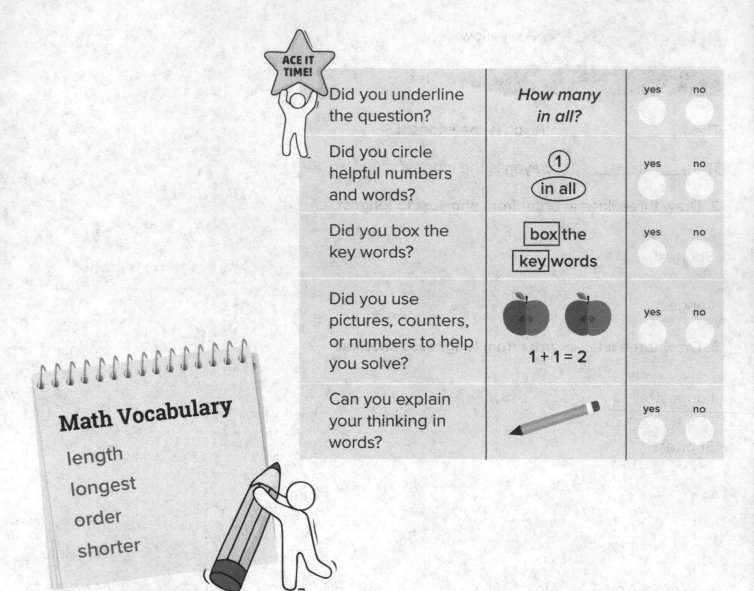

ACE IT TIME!

			yes	no
Did you underline the question?	*How many in all?*		○	○
Did you circle helpful numbers and words?	① in all		yes ○	no ○
Did you box the key words?	box the key words		yes ○	no ○
Did you use pictures, counters, or numbers to help you solve?	1 + 1 = 2		yes ○	no ○
Can you explain your thinking in words?			yes ○	no ○

Math Vocabulary

length

longest

order

shorter

Math on the Move

Go on a nature hike! Look for sticks or leaves outside your home. Put them in order from shortest to longest, and then from longest to shortest.

Use Non-Standard Units

FOLLOWING THE OBJECTIVE
You will measure objects using smaller objects.

LEARN IT: You can measure the length of an object by using smaller objects as a measuring tool. These smaller objects are called *units*. The units must all be the same size. There should be no spaces between the units.

Example: Use the square tiles to measure the line.

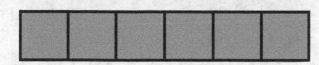

Length: about 6 square tile units.

think!
The line is about 6 square tile units long! Notice there are no spaces between the square tiles. Also, they are all the same size.

Example: If you only have one square tile, draw more square tiles to measure the line.

Length: about 4 square tile units.

87

PRACTICE: Now you try

Use the units shown to measure the objects.

1.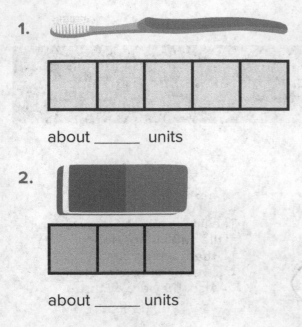

   about _____ units

2.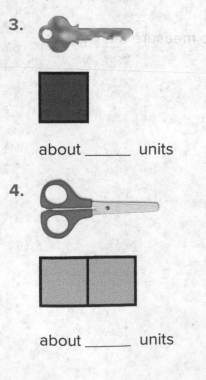

   about _____ units

How many of these square tile units would measure the objects below? Draw more units to help you.

3.

   about _____ units

4.

   about _____ units

Sam picked up this stick in his backyard. He said the stick was about 3 paper clips long. Do you agree? Why or why not? Explain your thinking. Show your work and write your explanation here.

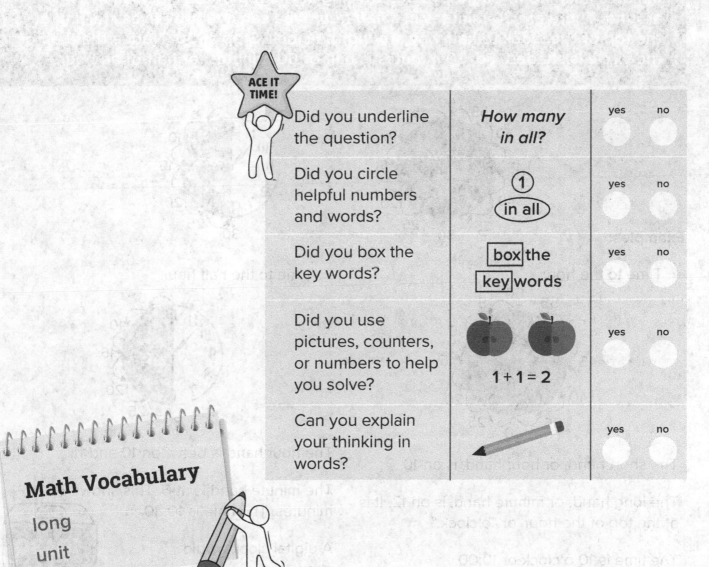

ACE IT TIME!		yes	no
Did you underline the question?	*How many in all?*	◯	◯
Did you circle helpful numbers and words?	① in all	◯	◯
Did you box the key words?	box the key words	◯	◯
Did you use pictures, counters, or numbers to help you solve?	1 + 1 = 2	◯	◯
Can you explain your thinking in words?		◯	◯

Math Vocabulary

long

unit

Math on the Move

Find some objects you would like to measure around your house. For the units, you will need small objects that are all the same size such as paper clips, buttons, or pennies. Practice measuring objects around the house using these units.

Tell Time

FOLLOWING THE OBJECTIVE
You will tell time to the hour and half hour.

LEARN IT: You can tell what time it is by reading a *clock*. There are 60 minutes in an hour. Count by fives starting at the top of the clock all the way around. There are 30 minutes in a *half hour*. You can count by fives starting at the top of the clock to half-way around the clock. Stop at 30. Notice that half of the circle is shaded. So, 30 minutes is half of 60 minutes, or half of an hour.

 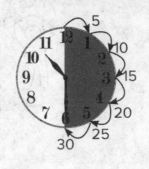

Examples:

1. Time to the hour.

The short hand, or hour hand, is on 10.

The long hand, or minute hand, is on 12. It is at the top of the hour, or "o'clock."

The time is 10 o'clock or 10:00.

2. Time to the half hour.

The hour hand is between 10 and 11.

The minute hand is at 6. This shows 30 minutes. The time is 10:30.

A digital clock would show this:

PRACTICE: Now you try

Read the analog clock. Write the time on the digital clock.

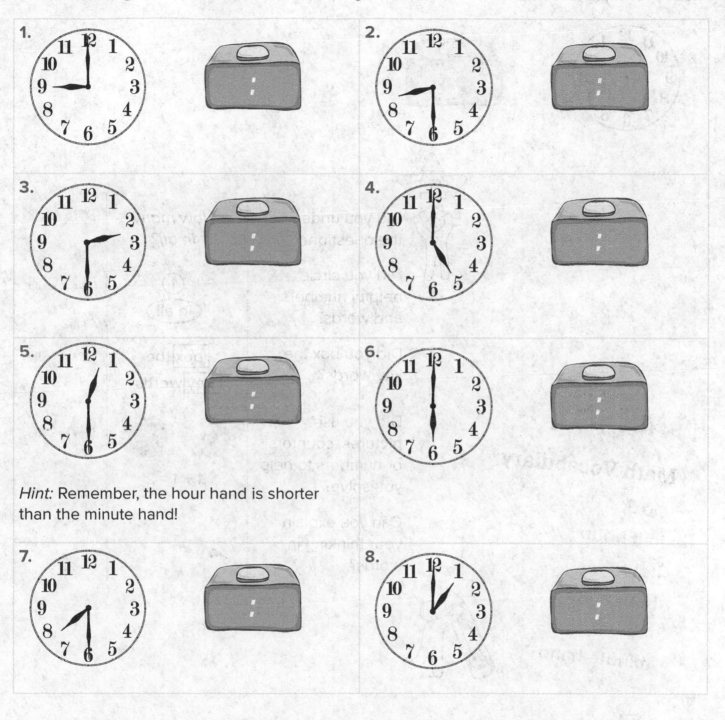

1.

2.

3.

4.

5.

Hint: Remember, the hour hand is shorter than the minute hand!

6.

7.

8.

Jack goes to bed at 8:00. His brother Tim goes to bed 30 minutes later. Draw and write what time Tim goes to bed on the analog and digital clocks below. How do you know? Show your work and write your explanation here.

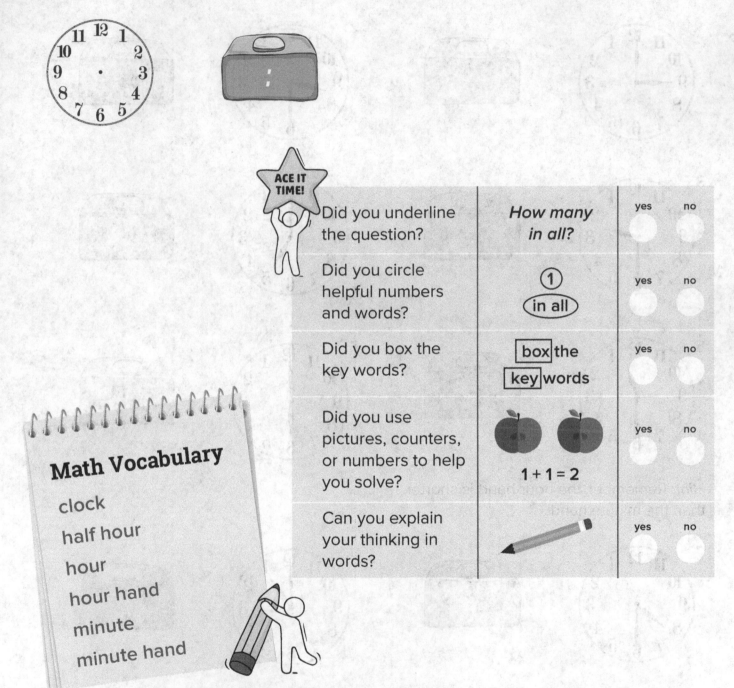

ACE IT TIME!	How many in all?	yes	no
Did you underline the question?		○	○
Did you circle helpful numbers and words?	① in all	○	○
Did you box the key words?	box the key words	○	○
Did you use pictures, counters, or numbers to help you solve?	1 + 1 = 2	○	○
Can you explain your thinking in words?		○	○

Math Vocabulary

clock

half hour

hour

hour hand

minute

minute hand

Math on the Move Watch the clock! Look at the clocks in your house. Some have hands (these are called analog clocks). Some just show the numbers (these are called digital clocks). Practice telling time on the hour and half hour using the clocks around your house.

Picture Graphs

FOLLOWING THE OBJECTIVE
You will read and answer questions about picture graphs.

LEARN IT: *Data* is another word for information. There are many different ways we can show data. A *picture graph* uses pictures to show data.

Example: The students in Mrs. Lee's class made a picture graph to show their favorite pets. Use the picture graph below to find more information.

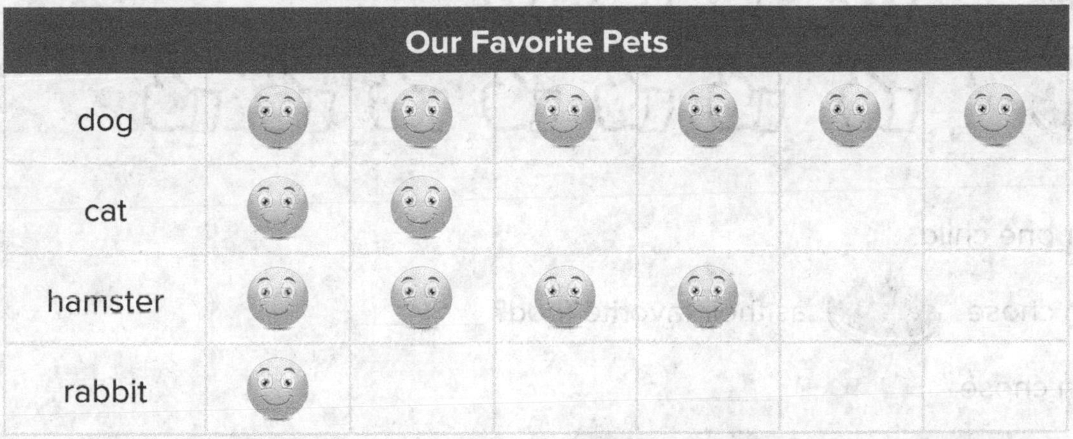

Our Favorite Pets						
dog	😊	😊	😊	😊	😊	😊
cat	😊	😊				
hamster	😊	😊	😊	😊		
rabbit	😊					

Each  😊 stands for one student.

How many students chose a dog as their favorite pet? *There are 6* 😊*. 6 students chose a dog as their favorite pet.*

Did more students choose a hamster or a dog? *There are more* 😊 *for dogs than hamsters. More students chose dogs than hamsters.*

How many more students chose a hamster than a rabbit? *Hint:* You have to compare the two! *There are 3 more* 😊 *for hamsters than rabbits.*

How many students chose a rabbit as their favorite pet? *There is 1* 😊*, so 1 student chose a rabbit as their favorite pet.*

PRACTICE: Now you try

Use the picture graph to answer the questions.

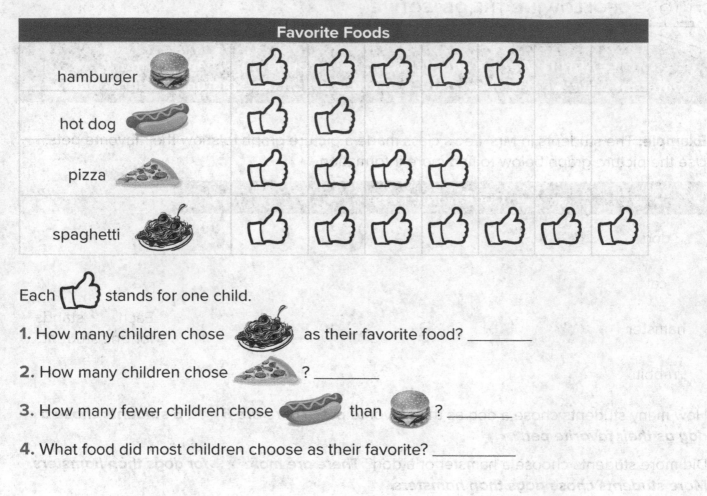

Each 👍 stands for one child.

1. How many children chose 🍝 as their favorite food? _____

2. How many children chose 🍕? _____

3. How many fewer children chose 🌭 than 🍔? _____

4. What food did most children choose as their favorite? _____

Use the favorite foods graph on page 94. How many more people chose  spaghetti than pizza and hot dogs combined? Explain your thinking. *Hint:* Subtract how many chose hot dogs and pizza from the total number of students that chose spaghetti. Show your work and write your explanation here.

Math Vocabulary

data
picture graph

	How many in all?	yes	no
ACE IT TIME! Did you underline the question?		○	○
Did you circle helpful numbers and words?	① in all	○	○
Did you box the key words?	box the key words	○	○
Did you use pictures, counters, or numbers to help you solve?	1 + 1 = 2	○	○
Can you explain your thinking in words?		○	○

Math on the Move

Keep track of the weather for one week. Make a picture graph to show whether it is sunny, cloudy, snowy, or rainy each day. Talk with your family members about the information you collected!

Bar Graphs and Tally Charts

FOLLOWING THE OBJECTIVE
You can read tally charts and bar graphs.

LEARN IT: Let's look at two more ways to show data. A *tally chart* uses *tally marks*. Tally marks are lines that show ones or groups of five. A *bar graph* uses bars of different lengths to show amounts. The longer the bar, the greater the amount.

Example: Jeni collected all the red, blue, and yellow crayons she could find in her house. The tally chart shows how many crayons of each color she found. She also made a bar graph with that same information.

think!
Jeni drew a tally mark for each crayon she found. Notice how 5 tallies are written like this: ⌦

Tally Chart

Crayons Jeni Found		Total
red 🖍️	I I I I	4
blue 🖍️	⌦ I I	7
yellow 🖍️	I I I	3

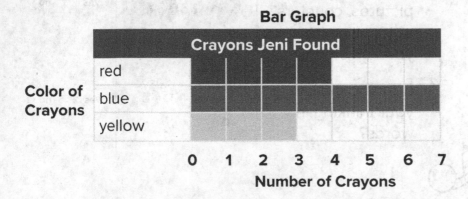

Bar Graph

think!
To read a bar graph, put your finger at the end of the bar. Trace the line down to find the number for that bar.

1. How many red crayons did Jeni find? _____

2. How many yellow crayons did Jeni find? _____

3. How many more blue crayons than yellow crayons did Jeni find? _____

4. What color crayon did Jeni find the most of? _____

5. What color crayon did Jeni find the least of? _____

6. How many crayons did Jeni find in all? _____

PRACTICE: Now you try

Use the bar graph to answer the questions.

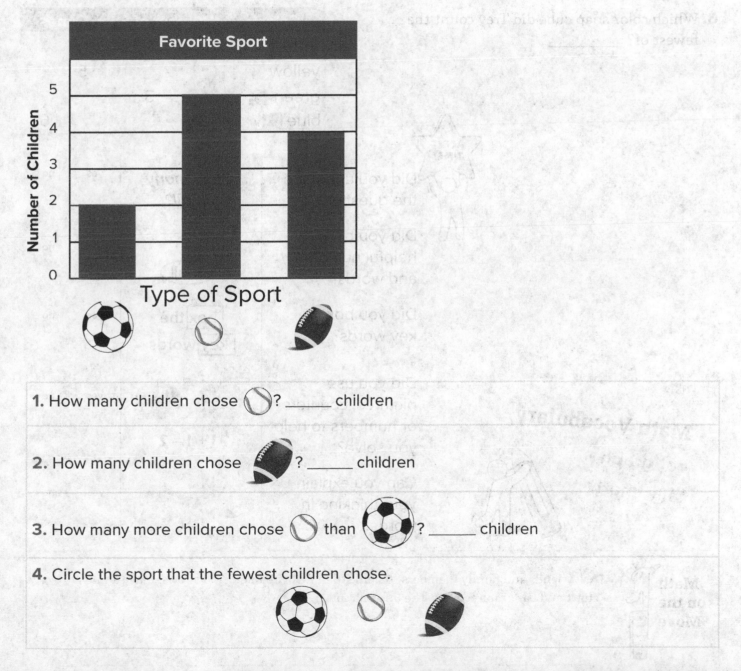

1. How many children chose 🏐? _____ children

2. How many children chose 🏈? _____ children

3. How many more children chose 🏐 than ⚽? _____ children

4. Circle the sport that the fewest children chose.

Unit 8: Measurement and Data Concepts

Trey counted some snap cubes. Use the information from his tally chart to make a bar graph. Then answer the questions.

1. How many ■ did Trey count? _____

2. How many ▧ did Trey count? _____

3. How many more ■ than ▧ did Trey count? _____

4. How many ■ and ▧ did Trey count? _____

5. Which color snap cube did Trey count the most of? _____

6. Which color snap cube did Trey count the fewest of? _____

Color		Number of Snap Cubes	Total
red	■	I I	2
yellow	▧	ＨＨＨ	5
green	■	I I I	3
blue	■	ＨＨＨ I	6

Color		Number of Snap Cubes					
red	■	2					
yellow	▧					5	
green	■		3				
blue	■						6

ACE IT TIME!

	How many in all?	yes	no
Did you underline the question?	*How many in all?*	○	○
Did you circle helpful numbers and words?	① in all	○	○
Did you box the key words?	box the key words	○	○
Did you use pictures, counters, or numbers to help you solve?	1 + 1 = 2	○	○
Can you explain your thinking in words?		○	○

Math Vocabulary

bar graph

tally chart

Math on the Move Ask friends and family members to name their favorite television shows. Make a tally chart and bar graph to show the information.

Geometry Concepts

Describe Two-Dimensional Shapes

FOLLOWING THE OBJECTIVE
You will be able to identify shapes by looking at how they are made.

LEARN IT: Some shapes are two-dimensional, or flat. You can describe and sort flat shapes by the number of *sides* and *vertices*, or corners they have and by their shapes.

A **circle** is a curve that is closed. It has no straight sides.

A **triangle** has 3 straight sides and 3 *vertices*.

A **rectangle** is a closed shape with 4 straight sides and 4 vertices. It also has 4 square corners.

A **square** is a closed shape with 4 straight sides that are the same length. It has 4 vertices. It has 4 square corners.

A **trapezoid** is a shape with 4 straight sides and 4 vertices. It has 2 sides that are opposite one another. They go in the same direction, but are different lengths.

Examples:

Circle the shapes that are triangles.

The first shape has 3 straight sides but it is not closed.

The third shape has 4 straight sides. It is a rectangle, not a triangle.

The second and last shapes are triangles.

Circle the shapes that are squares.

The first and second shapes are squares.

The third shape has two sides longer than the other two. It is a rectangle, not a square.

The last shape has no square corners. It is a trapezoid, not a square.

PRACTICE: Now you try

Write the name of the shape and draw it.

1. I am a shape with 3 sides and 3 vertices.

_____

2. I am a shape that is a closed curve.

_____

3. I am a shape that has 4 vertices and 4 sides, but not all the same length.

_____

4. I am a shape with 4 sides that are all the same length.

_____

5. I am a shape that has 4 sides, and 2 of them are parallel lines.

_____

Look at this picture of a house. Identify and count all the shapes you see in the picture. Show your work and write your explanation here.

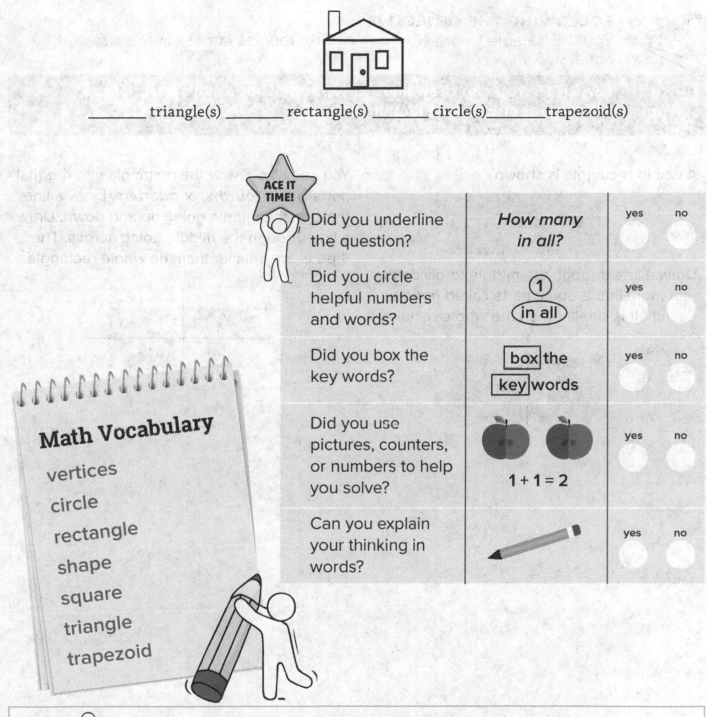

_____ triangle(s) _____ rectangle(s) _____ circle(s) _____ trapezoid(s)

ACE IT TIME!

	How many in all?	yes	no
Did you underline the question?		◯	◯
Did you circle helpful numbers and words?	① in all	◯	◯
Did you box the key words?	box the key words	◯	◯
Did you use pictures, counters, or numbers to help you solve?	1 + 1 = 2	◯	◯
Can you explain your thinking in words?		◯	◯

Math Vocabulary

vertices

circle

rectangle

shape

square

triangle

trapezoid

Math on the Move

Look around your house for squares, rectangles, circles, trapezoids, and triangles. Draw pictures of these items. Play a game of "I Spy" with the shapes you can find.

Circles and Rectangles: Halves and Fourths

FOLLOWING THE OBJECTIVE
You will be able to divide circles and rectangles into 2 equal parts or 4 equal parts.

LEARN IT: You can draw lines to separate shapes into *parts* of the *whole*. Let's look at circles and rectangles.

A whole rectangle is shown.

Draw a line through the middle to divide the rectangle into 2 equal parts called **halves**. Each half is smaller than the whole rectangle.

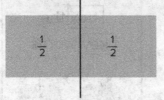

You can also divide the rectangle into 4 equal parts called **fourths**, or **quarters**. Draw a line through the middle going up and down. Draw a line through the middle going across. These 4 parts are smaller than the whole rectangle.

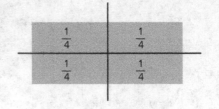

Examples:

Divide a circle into halves.

Draw a circle. Then draw a line through the middle.

$\frac{1}{2}$ | $\frac{1}{2}$

This circle is divided into 2 equal halves.

Divide the same circle into fourths.

There is already a line drawn down and through the middle. Draw a second line across and through the middle.

$\frac{1}{4}$ | $\frac{1}{4}$
$\frac{1}{4}$ | $\frac{1}{4}$

The circle is divided into 4 equal fourths.

PRACTICE: Now you try

1. Divide this circle into halves.

2. Divide this rectangle into fourths.

3. How much of this rectangle is striped?

_____

4. Circle the rectangle that shows fourths.

5. Divide the circle below into 4 equal parts.

6. Circle the shape that does not show halves.

7. Divide this candy bar into two equal pieces.

8. Divide this pie into enough pieces for 4 people to have an equal part.

Your class is having a pizza party. Would you rather have $\frac{1}{2}$ of a pizza or $\frac{1}{4}$ of a pizza? Draw a picture to show why. Show your work and write your explanation here.

ACE IT TIME!

		yes	no
Did you underline the question?		○	○
Did you circle helpful numbers and words?	① in all	○	○
Did you box the key words?	box the key words	○	○
Did you use pictures, counters, or numbers to help you solve?	🍎 🍎 1 + 1 = 2	○	○
Can you explain your thinking in words?		○	○

How many in all?

Math Vocabulary

fourths

halves

quarters

whole

Math on the Move

Practice folding a piece of paper in half. Can you fold it again to make fourths? Try cutting out circles to fold into halves or fourths.

Describe Three-Dimensional Shapes

FOLLOWING THE OBJECTIVE
You will describe three-dimensional shapes.

LEARN IT: Shapes can be sorted in different ways. They can be sorted by the number of *sides* and vertices, or corners. Two-dimensional shapes are flat. ***Three-dimensional shapes*** are solid. They may be sorted by their ***curved*** and ***flat surfaces***.

		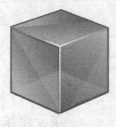		
A ***sphere*** has 0 flat surfaces and a curved surface. It can roll.	A ***cone*** has 1 flat surface that is a circle. It also has 1 curved surface. A cone can roll on its side.	A ***cube*** has 6 flat surfaces that are squares. You can stack it. It has 0 curved surfaces.	A ***cylinder*** has 2 flat surfaces that are circles. It also has 1 curved surface. A cylinder can stack and roll.	A ***rectangular prism*** has 6 flat surfaces that are rectangles. You can stack it.

PRACTICE: Now you try

Write the name of the three-dimensional shape that each object is most like and fill in the blanks.

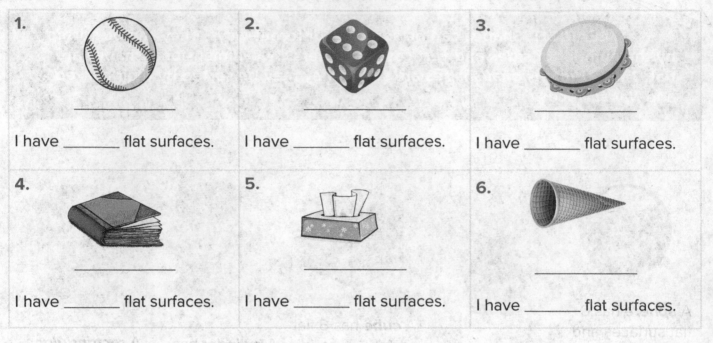

1. _____

   I have _____ flat surfaces.

2. _____

   I have _____ flat surfaces.

3. _____

   I have _____ flat surfaces.

4. _____

   I have _____ flat surfaces.

5. _____

   I have _____ flat surfaces.

6. _____

   I have _____ flat surfaces.

Rachel solved the following problem: "Think of a shape that has flat surfaces and curved surfaces. You can stack it. Name the shape and draw a picture of it." See Rachel's work below. Did she solve the problem correctly? How do you know? Show your work and explain your thinking.

Rachel's work: cube

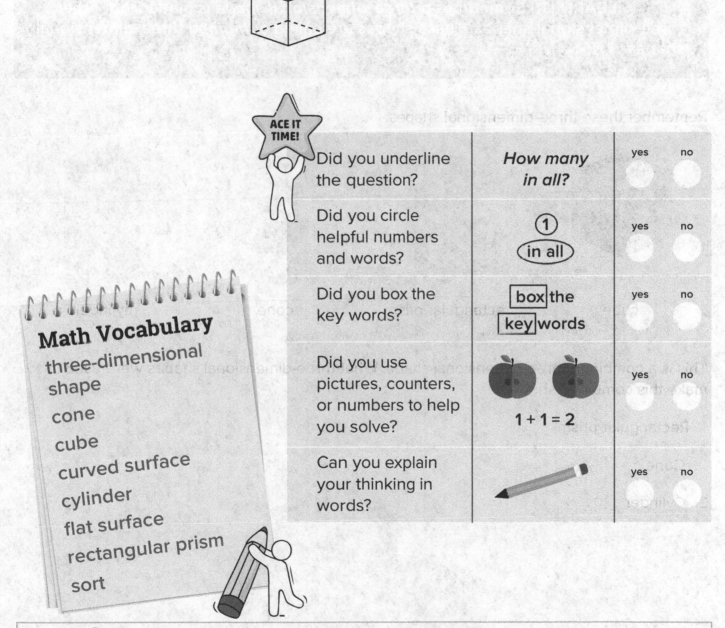

Math Vocabulary

three-dimensional shape

cone

cube

curved surface

cylinder

flat surface

rectangular prism

sort

ACE IT TIME!			yes	no
Did you underline the question?	*How many in all?*		◯	◯
Did you circle helpful numbers and words?	① in all		◯	◯
Did you box the key words?	box the key words		◯	◯
Did you use pictures, counters, or numbers to help you solve?	1 + 1 = 2		◯	◯
Can you explain your thinking in words?			◯	◯

Math on the Move — Look around the house for two-dimensional shapes. What shape is a can? What shape is a globe?

Combine Three-Dimensional Shapes

FOLLOWING THE OBJECTIVE
You will be able to make combined three-dimensional shapes from other three-dimensional shapes.

LEARN IT: Sometimes a *three-dimensional shape* is made from other three-dimensional shapes. These shapes are called *combined* three-dimensional shapes.

Remember these three-dimensional shapes.

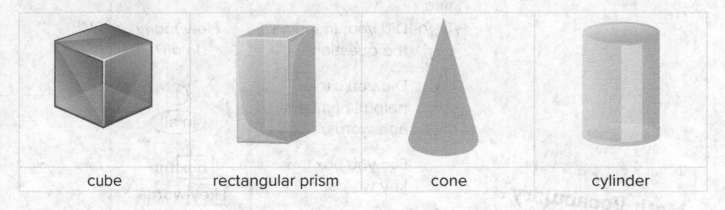

| cube | rectangular prism | cone | cylinder |

This is a combined three-dimensional shape. What three-dimensional shapes were used to make this combined shape?

1. Rectangular prism

2. Cone

3. Cylinder

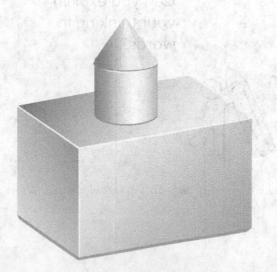

PRACTICE: Now you try

Answer the questions.

1. What simple three-dimensional shapes are in this combined shape?

_____

_____

2. What simple three-dimensional shapes are in this combined shape?

_____

_____

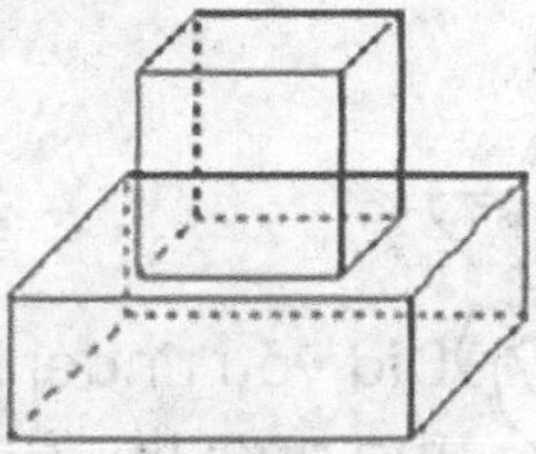

3. What simple three-dimensional shapes are in this combined shape?

_____

_____

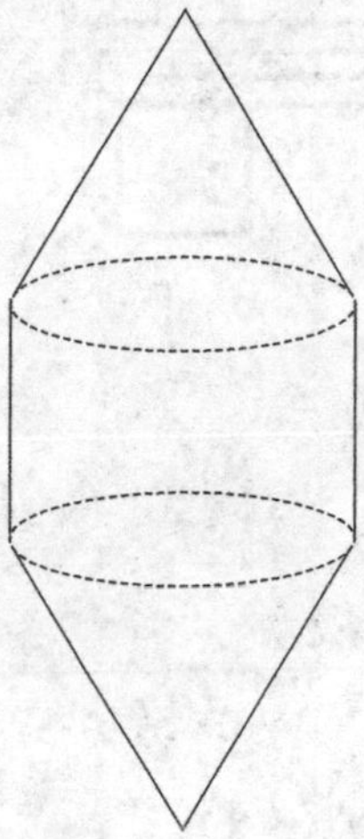

think!
This is made up of 3 different figures.

Kristina used blocks to build part of a tower. How many blocks did she use? What are the shapes of the blocks? Show your work and explain your thinking.

Kristina's Figure

		yes	no
ACE IT TIME! Did you underline the question?	*How many in all?*	yes ○	no ○
Did you circle helpful numbers and words?	① in all	yes ○	no ○
Did you box the key words?	box the key words	yes ○	no ○
Did you use pictures, counters, or numbers to help you solve?	1 + 1 = 2	yes ○	no ○
Can you explain your thinking in words?		yes ○	no ○

Math Vocabulary

combined shape

Math on the Move

Work with a partner. Write a list of three-dimensional shapes and give it to your partner. Have your partner draw a combined shape made of these three-dimensional shapes. Take turns drawing!

REVIEW

Congratulations! You have finished the lessons for Units 8 and 9. That means you have learned how to order and compare lengths with non-standard units. You can tell time to the hour and half hour. You can also show data in tally charts, picture graphs, and bar graphs. You also can describe two- and three-dimensional shapes, and combine three-dimensional shapes to make other shapes.

Now it's time to show your skills. Solve the problems below using what you have learned.

Activity Section 1

1. Draw three lines from shortest to longest.

Shortest	
Longest	

2. How long is this line?

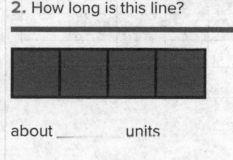

about _____ units

3. What time does the clock show?

4. Draw hands on the clock to match the time.

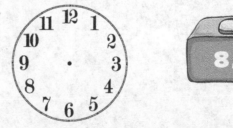

5. Read the picture graph below. Answer the questions.

How Do You Get to School?

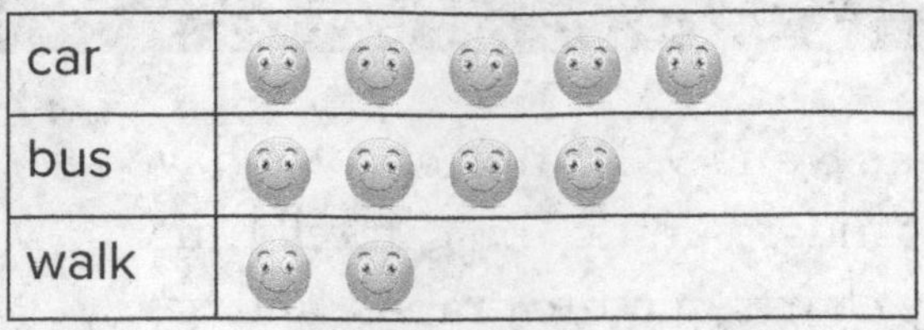

car	😊 😊 😊 😊 😊
bus	😊 😊 😊 😊
walk	😊 😊

Each 😊 stands for one child.

a. How many children ride the bus to school?
_____

b. How many children walk to school?
_____

c. How many fewer children walk than take the bus to school? _____

6. Read the bar graph. Answer the questions.

Favorite Color Jelly Bean

pink					
purple					
red					
orange					

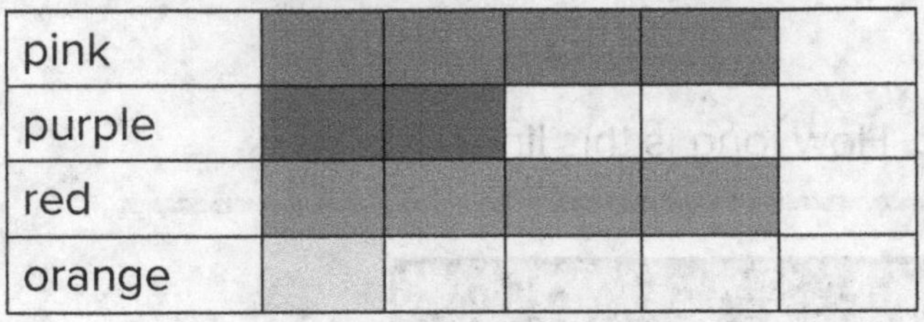

a. How many students chose purple as their favorite color jelly bean? _____

b. How many students chose purple and orange as their favorite color jelly bean?
_____

c. How many more chose pink than purple as their favorite color jelly bean? _____

d. Which colors of jelly beans were chosen by an equal number of students?
_____

Activity Section 2

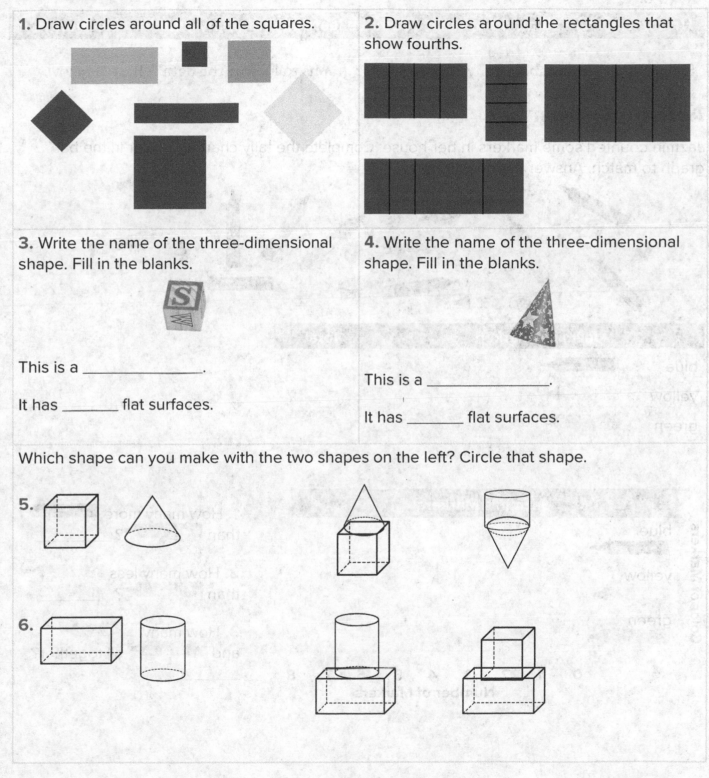

1. Draw circles around all of the squares.

2. Draw circles around the rectangles that show fourths.

3. Write the name of the three-dimensional shape. Fill in the blanks.

This is a _____.

It has _____ flat surfaces.

4. Write the name of the three-dimensional shape. Fill in the blanks.

This is a _____.

It has _____ flat surfaces.

Which shape can you make with the two shapes on the left? Circle that shape.

5.

6.

UNDERSTAND

Understand the meaning of what you have learned and apply your knowledge.

Use what you know about tally charts and bar graphs to look at the data collected below.

Activity Section

Jazmin counted some markers in her house. Complete the tally chart and color in the bar graph to match. Answer the questions.

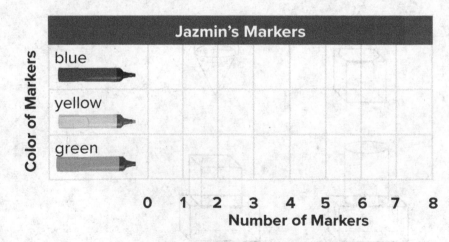

Jazmin's Markers		Total
blue		
yellow		
green		

Jazmin's Markers								
blue								
yellow								
green								

Color of Markers

0 1 2 3 4 5 6 7 8
Number of Markers

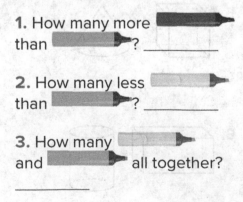

1. How many more than ? _____

2. How many less than ? _____

3. How many and all together? _____

114

DISCOVER

Two- and three-dimensional shapes are all around us! Think of ways you use shapes in your everyday life.

Activity Section

Design a party hat. What three-dimensional shape would you use? Draw your hat below. Decorate it with other two-dimensional shapes for a party! Label the shapes on your hat below.

Answer Key

Unit 2: Addition Concepts

Use Pictures to Add

Page 10 Practice: Now you try

1. 7, 7
2. 4, 4
3. 6, 6
4. 7, 7
5. 9, check student's drawing
6. 8, check student's drawing
7. 6, check student's drawing
8. 5, check student's drawing

Page 11 Ace It Time: 8 birds; check student's drawing; 5 + 3 = 8; I added 5 + 3 to get 8.

Use Counters to Add

Page 13 Practice: Now you try

1. 3, check student's drawing
2. 8, check student's drawing
3. 7, check student's drawing
4. 7, check student's drawing
5. check student's drawing, 6 bunnies, 3 + 3 = 6
6. check student's drawing, 9 grapes, 3 + 6 = 9
7. check student's drawing, 10 children, 8 + 2 = 10
8. check student's drawing, 10 trucks, 9 + 1 = 10

Page 14 Ace It Time: 9 apples; 4 + 5 = 9; I counted 4 apples and 5 apples to get 9.

Add in Any Order

Page 16 Practice: Now you try

1. 6, 4 + 2 = 6, check student's coloring
2. 8, 5 + 3 = 8, check student's coloring
3. 7, 2 + 5 = 7, check student's coloring
4. 9, 3 + 6 = 9, check student's coloring

Page 17 Ace It Time: They are both correct. 3 + 7 and 7 + 3 give the same sum of 10.

Unit 3: Subtraction Concepts

Use Pictures to Subtract

Page 19 Practice: Now you try

1. 1, 1
2. 1, 1
3. 2, 2
4. 6, 6
5. 6, check student's drawing
6. 4, check student's drawing
7. 3, check student's drawing
8. 2, check student's drawing

Page 20 Ace It Time: 4 balloons, check student's drawing, 9 – 5 = 4; I drew 9 circles and crossed out 5 to get 4.

Use Counters to Subtract

Page 22 Practice: Now you try

1. 1, check student's drawing
2. 3, check student's drawing
3. 4, check student's drawing
4. 2, check student's drawing
5. Check student's drawing, 6 – 3 = 3 red ants
6. Check student's drawing, 7 – 3 = 4 girls
7. Check student's drawing, 10 – 7 = 3 cats with no stripes
8. Check student's drawing, 9 – 2 = 7 short trees

Page 23 Ace It Time: 8 geese, 10 – 2 = 8; I made a pile of 10 counters. I moved 2 away, and 8 were left.

Subtract and Compare

Page 25 Practice: Now you try

1. 2, 2
2. 6, 6
3. 7, 9 – 2 = 7
4. 2, 7 – 5 = 2

Page 26 Ace It Time: 3 fewer stickers, 10 – 7 = 3, I made a row of 7 counters for Mike. I made a row of 10 counters for Leah. Three did not line up.

Stop and Think! Units 2–3 Review

Page 27 Activity Section 1

1. 4, 4
2. Check student's drawing, 5
3. Check student's drawing, 6
4. 10 hats, 8 + 2 = 10

Page 28

5. 9, 4 + 5 = 9, check student's coloring
6. 8 dogs, 2 + 6 = 8, 6 + 2 = 8

Activity Section 2

1. 4, 4 apples left
2. Check student's drawing, 7 – 3 = 4, 4 bunnies left
3. Check student's drawing, 7
4. 6 small dogs

Page 29

5. 2, 2 more squirrels
6 10 – 8 = 2, 2 fewer markers

Activity Section 3

1. 9
2. 5
3. 2
4. 6
5. 4
6. 10
7. 4
8. 5
9. 3

Stop and Think! Units 2–3 Understand

Page 30

Both children wrote correct subtraction sentences, but Abby's work is wrong; Abby drew 7 gumballs. She crossed out 3. She wrote 7 – 4 = 3. Her drawing shows 7 – 3 = 4.

Nina's work is right. She drew 7 gumballs and 3 gumballs in rows. She lined up 3 pairs of gumballs. 4 were not lined up. She wrote 7 – 3 = 4.

Stop and Think! Units 2–3 Discover

Page 31

$3: pinwheel and pencil, $2 + $1 = $3, $10 – $3 = $7; $6: pinwheel and car, $2 + $4 = $6, $10 – $6 = $4; $10: pinwheel and ball, $2 + $8 = $10, or sunglasses and pencil, $9 + $1 = $10, $10 – $10 = $0.

Unit 4: Addition and Subtraction Strategies

Use a Double Ten Frame to Add

Page 33 Practice: Now you try

1. 15, 10 + 5 = 15, check student's drawing
2. 12, 10 + 2 = 12, check student's drawing
3. 16, 10 + 6 = 16, check student's drawing

Page 34 Ace It Time: 11 toys; Possible answer: I put 9 counters in the top ten frame. I put 2 counters in the bottom ten frame. I moved 1 counter up to make 10. There are 10 in the top frame and 1 in the bottom. The sum is 11. I added 9 + 2 instead of 2 + 9, because it is easier to move 1 counter to make 10 than to move 8 counters to make 10.

Count On to Add

Page 36 Practice: Now you try

1. 10; 6 is circled
2. 12; 8 is circled
3. 11; 9 is circled
4. 10; 7 is circled
5. 13; 9 is circled
6. 17; 9 is circled
7. 15; 9 is circled
8. 13; 8 is circled

Page 37 Ace It Time: Marco and Tayo both counted on from one of the addends and got the correct sum, 12. Possible answer: I like Tayo's way better. He counted on from the greater addend. So, he had less counting to do.

Use a Double Ten Frame to Subtract

Page 39 Practice: Now you try

1. 4, check student's drawing
2. 3, check student's drawing
3. 5, check student's drawing
4. 8, check student's drawing

Page 40 Ace It Time: 7 beads are blue, 16 – 9 = 7, check student's drawing.

Use Addition to Subtract

Page 42 Practice: Now you try

1. 2, 8 + 2 = 10, 10 – 8 = 2
2. 6, 6 + 6 = 12, 12 – 6 = 6
3. 4, 7 + 4 = 11, 11 – 7 = 4
4. 8, 5 + 8 = 13, 13 – 5 = 8
5. 10, 4 + 10 = 14, 14 – 4 = 10

Page 43 Ace It Time: 8 blue butterflies, 6 + 8 = 14, so 14 – 6 = 8.

Add Three Numbers

Page 45 Practice: Now you try

1. 14, 2 + 4 = 6, 6 + 8 = 14, 14, 2 + 8 = 10, 10 + 4 = 14
2. 12, 7 + 2 = 9, 9 + 3 = 12, 12, 7 + 3 = 10, 10 + 2 = 12
3. 3 + 4 = 7, 7 + 3 = 10 or 4 + 3 = 7, 7 + 3 = 10 or 3 + 3 = 6, 6 + 4 = 10
4. 9 + 0 = 9, 9 + 1 = 10 or 0 + 1 = 1, 1 + 9 = 10 or 9 + 1 = 10, 10 + 0 = 10
5. 2 + 6 = 8, 8 + 2 = 10 or 2 + 2 = 4, 4 + 6 = 10
6. 5 + 7 = 12, 12 + 5 = 17 or 5 + 5 = 10, 10 + 7 = 17

Page 46 Ace It Time: 15 apples; check student's drawing. For grouping, explanations may include: I added 3 + 7 to make 10, then I added 10 + 5 = 15, or I added 5 + 7 to make 12, then I added 12 + 3 = 15, or I added 3 + 5 to make 8, then I added 8 + 7 = 15.

Answer Key

Unit 5: Relationships with Operations
Find Missing Numbers
Page 48 Practice: Now you try
1. 8, 8
2. 5, 5
3. 7, 7
4. 4, 4
5. 5, 5
6. 5, 5
7. 6, 6
8. 9, 9

Page 49 Ace It Time: 5 + 2 + ? = 13 OR 2 + 5 + ? = 13; ? = 6 gray marbles. Possible explanation: I added 5 blue marbles + 2 yellow marbles = 7 marbles that are not gray. I subtracted 13 total marbles − 7 blue and yellow marbles = 6 gray marbles.

Choose an Operation
Page 51 Practice: Now you try
1. 11 − 5 = 6, 6 crackers
2. 8 + 7 = 15, 15 pencils
3. 14 − 5 = 9 or 5 + 9 = 14, 9 trucks
4. 12 + 4 = 16, 16 stuffed bears

Page 52 Ace It Time: 10 cards; Students might add 8 + ? = 18 and get the answer. They might also subtract 18 − 8 = ? to get the answer. Whichever method students use, they should mention the other method as a possible way.

Equal or Not Equal
Page 54 Practice: Now you try
1. False
2. True
3. True
4. True
5. False
6. True
7. True
8. False
9. True

Page 55 Ace It Time: Andrea and Paula both counted 10 stars. They are the same. 5 + 3 + 5 is equal to 5 + 5 + 3 = 10 + 3.

Stop and Think! Units 4–5 Review
Page 56 Activity Section
1. 13, 2. 10 + 3 = 13, check student's drawing

Page 57
3. 3, check student's drawing
4. 5 + 6 = 11, 6

5. 11, 7 should be circled
6. 14, 9 should be circled
7. 12, 6 + 3 = 9, 9 + 3 = 12 or 3 + 3 = 6, 6 + 6 = 12
8. 18, 8 + 4 = 12, 12 + 6 = 18 or 4 + 6 = 10, 10 + 8 = 18 or 8 + 6 = 14, 14 + 4 = 18. For questions 6–7, choice of which addends to add first will vary. Check that sums in boxes are correct for whichever addends students chose.

Page 58
9. 7, 7
10. 8, 8
11. 9 + 6 = 15, 15 deer
12. 19 − 9 = 10, 10 orange fish
13. False
14. False
15. False
16. True

Stop and Think! Units 4–5 Understand
Page 59 Activity Section

No, it is not the only way. Other methods will vary. Possible answer: She could write an addition sentence instead of a subtraction one, 9 + [__] = 15. She could use a related fact, 15 − 9 = 6, so 9 + 6 = 15. The answer is still 6.

Stop and Think! Units 4–5 Discover
Page 60 Activity Section

15 − 7 = 8 (or 8 + 7 = 15); 8 children eat sandwiches or soup; possible numbers will vary but should have a sum of 8. For example, 5 eat sandwiches and 3 eat soup.

Unit 6: Number Concepts
Count by Ones and Tens to 120
Page 62 Practice: Now you try
1. 11, 12, 13, 14
2. 55, 65, 75, 85
3. 43, 53, 63, 73
4. 105, 106, 107, 108
5. 92, 93, 94, 95
6. 20, 21, 22, 23
7. 72, 82, 92, 102
8. 59, 60, 61, 62

Page 63 Ace It Time: 50 coins; write the series 10, 20, 30, 40, 50 to count by tens

Tens and Ones Through 120
Page 65 Practice: Now you try
1. 1 ten, 7 ones
2. 8 tens, 3 ones
3. 5 tens, 5 ones

4. 3 tens, 2 ones

5. 7 tens, 9 ones

6. 9 tens, 0 ones

7. 6 tens, 4 ones

8. 1 ten, 1 one

Page 66 Ace It Time: 26; check student's drawing.

Compare Numbers Using <, >, =

Page 68 Practice: Now you try

1. >

2. >

3. =

4. <

5. <

6. >

7. >

8. =

Page 69 Ace It Time: 43 > 39 or 39 < 43; Carlos; check student's drawing.

Unit 7: Two-Digit Addition and Subtraction Concepts

Add and Subtract Tens

Page 71 Practice: Now you try

1. 80

2. 70

3. 0

4. 10

5. 10

6. 80

7. 20

8. 60

Page 72 Ace It Time: 60 jumping jacks; 40 + 20 = 60; 4 tens + 2 tens is 6 tens.

Mental Math: Add 10 or Subtract 10

Page 74 Practice: Now you try

1. 45

2. 41

3. 89

4. 54

5. 38

6. 92

7. 20

8. 89

Page 75 Ace It Time: 33 students; 43 – 10 = 33

Use Tens and Ones to Add

Page 77 Practice: Now you try

1. 15

2. 36

3. 53

4. 68

5. 80

6. 82

7. 82

8. 66

Page 78 Ace It Time: 76 points; 35 + 41 = 76

Stop and Think! Units 6–7 Review

Page 79 Acitivty Section 1

1. 116, 117, 118

2. 41, 42, 43

3. 13, 14, 15

4. 88, 89, 90

5. 38, 48, 58

6. 56, 66, 76

7. 85, 95, 105

8. 60, 70, 80

Page 80 Activity Section 2

1. 8 tens, 0 ones

2. 2 tens, 9 ones

3. 3 tens, 7 ones

4. 9 tens, 3 ones

5. 1 ten, 2 ones

6. 7 tens, 2 ones

7. 4 tens, 5 ones

8. 6 tens, 8 ones

Page 80 Activity Section 3

1. >

2. =

3. <

4. <

5. =

6. <

7. <

8. >

Page 81 Activity Section 4

1. 20

2. 40

3. 60

4. 10

5. 0

6. 30

7. 40

8. 80

Page 81 Activity Section 5

1. 51

2. 26

3. 75

4. 63

5. 12

6. 29

7. 78

8. 69

Page 81 Activity Section 6

1. 61

2. 98

3. 33

4. 90

5. 90

6. 81

7. 97

8. 84

Stop and Think! Units 6–7 Understand

Page 82 Activity Section

Cameron saved $42. After he buys the game, Cameron has $22. Possible explanation: 4 tens + 2 ones = 42; 20 = 2 tens; 42 − 20 = 22.

Stop and Think! Units 6–7 Discover

Page 83 Activity Section

51 students are going on the field trip. Yes, the students fit on the bus. Possible explanation: 27 + 24 = 51 students in all. 51 students < 56 bus seats.

Unit 8: Measurement and Data Concepts

Order and Compare Lengths

Page 85 Practice: Now you try

1. Blue, red

2. Check student's drawing

3. Check student's drawing

Page 86 Ace It Time: The blue pencil is the longest. The orange pencil is the shortest, and the green pencil is in the middle. I know because I drew a picture.

Use Non-Standard Units

Page 88 Practice: Now you try

1. About 5 units

2. About 3 units

3. About 2 units

4. About 3 units

Page 89 Ace It Time: Sam is not correct. The paper clip units should not have any spaces in between. The stick is actually about 5 paper clips long.

Tell Time

Page 91 Practice: Now you try

1. 9:00

2. 8:30

3. 2:30

4. 5:00

5. 12:30

6. 6:00

7. 7:30

8. 1:00

Page 92 Ace It Time: Tim went to bed at 8:30. I know because that is 30 minutes, or a half hour, past 8:00. The short hand, or hour hand, is on the 8. The long hand, or minute hand, is on the 6 to show 30 minutes.

Picture Graphs

Page 94 Practice: Now you try

1. 7

2. 4

3. 3

4. Spaghetti

Page 95 Ace It Time: One more student chose spaghetti than pizza and hot dog combined. Possible explanation: 4 people chose pizza and 2 people chose hot dogs. 4 + 2 = 6. I subtracted that from the number of spaghetti choices. 7 − 6 = 1.

Bar Graphs and Tally Charts

Page 96

Example:

1. 4

2. 3

Page 97

3. 4

4. Blue

5. Yellow

6. 14 crayons, 4 + 7 + 3 = 14

Practice: Now you try

1. 5

2. 4

3. 3

4. Soccer had the fewest.

Page 98 Ace It Time:

1. 6

2. 5

3. 4

4. 5

5. Blue had the most

6. Red had the fewest

Color	Number of Snap Cubes				
red					
yellow					
green					
blue					

Unit 9: Geometry Concepts

Describe Two-Dimensional Shapes

Page 100 Practice: Now you try

1. Triangle, check student's drawing

2. Circle, check student's drawing

3. Rectangle, check student's drawing

4. Square, check student's drawing

5. Trapezoid, check student's drawing

Page 101 Ace It Time: 1 triangle, 4 rectangles, 1 circle.

Circles and Rectangles: Halves and Fourths

Page 103 Practice: Now you try

1. Circle should be divided into halves using one line

2. Rectangle should be divided into fourths using two lines

3. $\frac{1}{2}$

4. The rectangle divided into fourths should be circled

5. Circle should be divided into fourths using two lines

6. The rectangle on the second row should be circled

7. Candy bar should be divided into halves using one line

8. Pie should be divided into fourths using two lines

Page 104 Ace It Time: Picture should represent a pizza/circle cut into halves and a pizza/circle cut into fourths. Possible answer: I would rather have $\frac{1}{2}$ of a pizza because it is more than $\frac{1}{4}$ of a pizza.

Describe Three-Dimensional Shapes

Page 106 Practice: Now you try

1. Sphere, 0

2. Cube, 6

3. Cylinder, 2

4. Rectangular prism, 6

5. Rectangular prism, 6

6. Cone, 1

Page 107 Ace It Time: Rachel is not correct. The cube shape she drew does not have any curved surfaces. The correct answer is a cylinder, because it has flat surfaces, curved surfaces, and it can be stacked.

Combine Three-Dimensional Shapes

Page 109 Practice: Now you try

1. rectangular prism, cylinder

2. cube, rectangular prism

3. cylinder, 2 cones

Page 110 Ace It Time: 6 blocks; 3 cones, 1 rectangular prism, 2 cubes

Stop and Think! Units 8–9 Review

Page 111 Activity Section 1

1. Answers will vary, check student's drawing

2. About 6 units

3. 4:00

4. 8:30, check clock hands

Page 112

5. a. 4; b. 2; c. 2

6. a. 2; b. 3; c. 2; d. pink and red

Page 113 Activity Section 2

1.

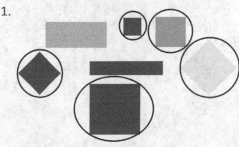

2. Student should circle the top 3 rectangles that show equal fourths

3. Cube; It has 6 flat surfaces

4. Cone; It has 1 flat surface

5.

Answer Key

6.

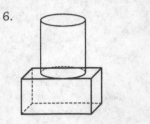

Stop and Think! Units 8–9 Understand
Page 114 Activity Section

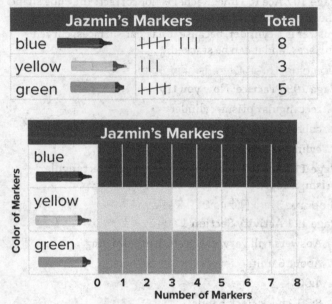

Jazmin's Markers		Total
blue	⊮⊮⊮ III	8
yellow	III	3
green	⊮⊮⊮	5

1. 3
2. 2
3. 8

Stop and Think! Units 8–9 Discover
Page 115 Activity Section
Answers will vary. Students should draw a cone shape for a party hat. They can use any other 2-D shapes to decorate the hat.

120s Chart

1	2	3	4	5	6	7	8	9	10
11	12	13	14	15	16	17	18	19	20
21	22	23	24	25	26	27	28	29	30
31	32	33	34	35	36	37	38	39	40
41	42	43	44	45	46	47	48	49	50
51	52	53	54	55	56	57	58	59	60
61	62	63	64	65	66	67	68	69	70
71	72	73	74	75	76	77	78	79	80
81	82	83	84	85	86	87	88	89	90
91	92	93	94	95	96	97	98	99	100
101	102	103	104	105	106	107	108	109	110
111	112	113	114	115	116	117	118	119	120

Telling Time Practice

Directions: Choose a time of day. Add the hour and minute hands to the clocks below. Remember, the hour hand is shorter than the minute hand. Write in the same time on the digital clock.

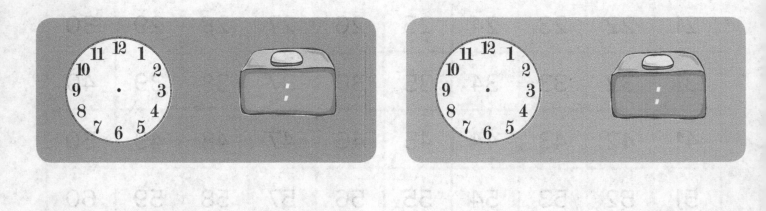

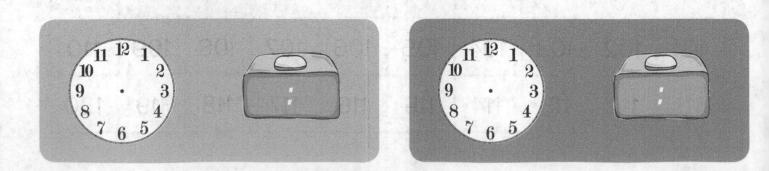

Cut Out Counters

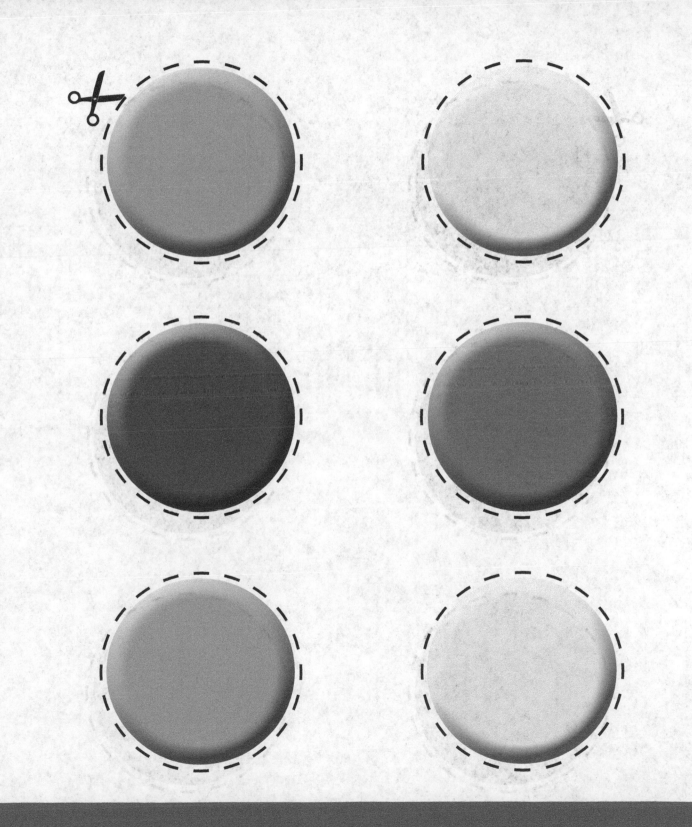

Cut Out Counters

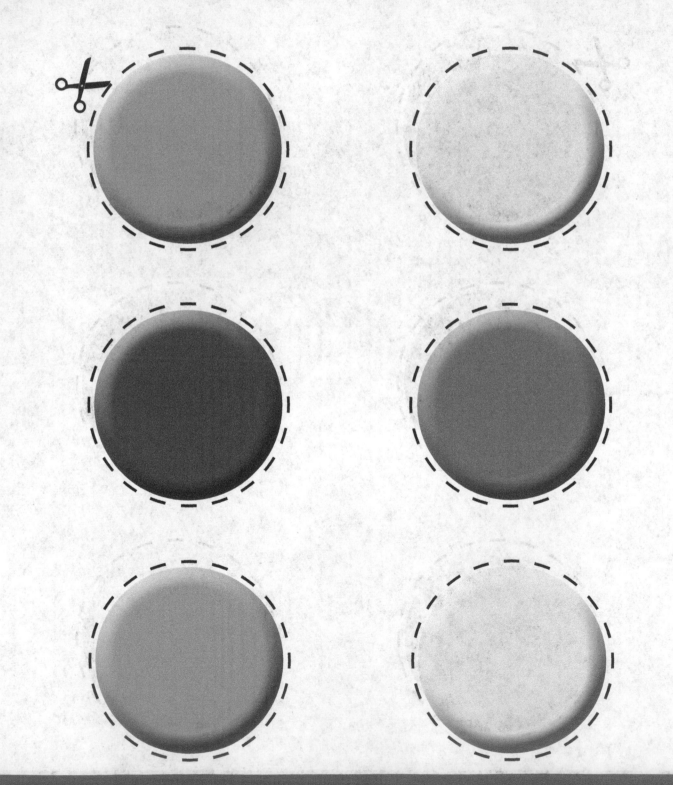

Double Ten Frame Workmat

Double Ten Frame Workmat

Double Ten Frame Workmat

Bar Graphs and Tally Charts

Ask your friends and family which three-dimensional shape is their favorite. Complete the tally chart and the bar graph. Record your data below!

Tally Chart

The data I have tells me that ...

Bar Graph

15				
14				
13				
12				
11				
10				
9				
8				
7				
6				
5				
4				
3				
2				
1				

Measure It

Directions: Pick a non-standard measurement tool (see the next page for examples) to estimate and measure five objects in your home. Show your work below.

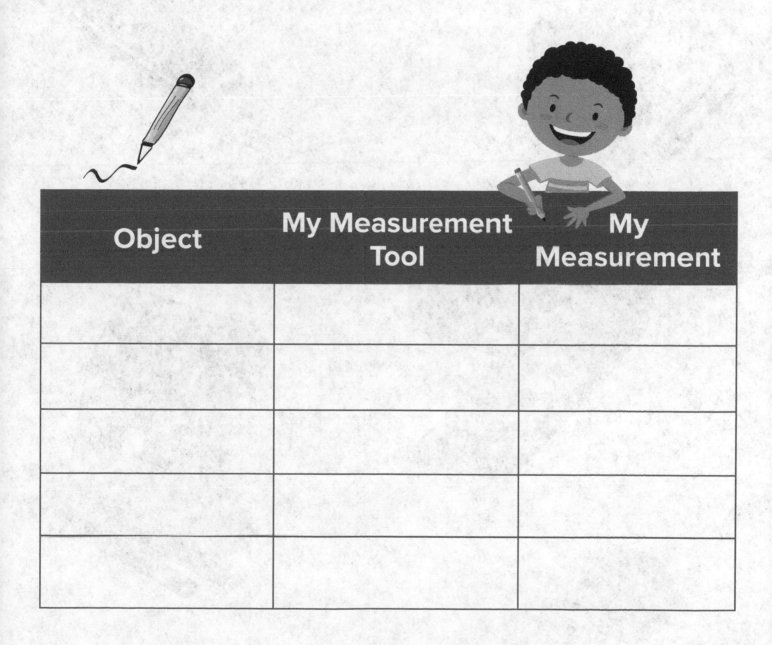

Object	My Measurement Tool	My Measurement

Non-Standard Measurement Objects

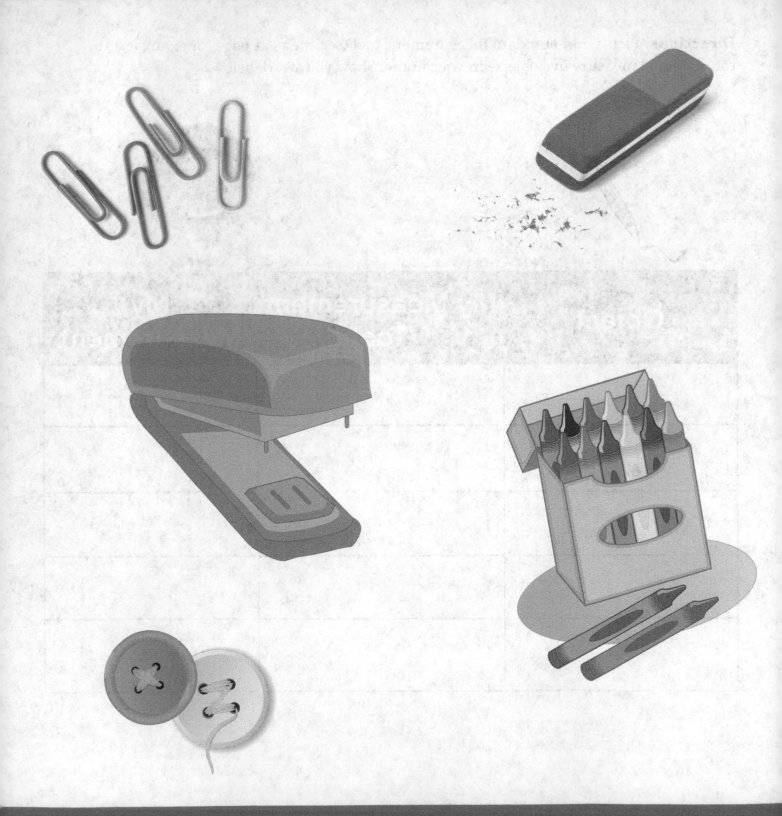